ÆSTHETIC NOMADS

PHOTOGRAPHY & DESIGN: REINHILDE GIELEN
WORDS: HANS PAUWELS

ÆSTHETIC NOMADS

REINHILDE GIELEN & HANS PAUWELS

BETA-PLUS

TABLE OF CONTENTS

The Art of Exp
Charlie Trotter's Desserts
Erwin Olaf
HENRY VAN DE VELDE
CHARLOTTE PERRIAND

Some people know what their lives will be like. From their earliest childhood, they want to become dentists, pilots, teachers, or lawyers. They devote their very existence to accomplishing just such an objective. We've never felt like that; we had no idea what we wanted, but we were sure that life for us would take place elsewhere. The irresistible appeal of the unknown—which, at that time, meant whatever lay beyond the limits of the rather provincial towns we grew up in—affected us both. We grasped at every opportunity to make this come about—soulmates long before we met.

We each conceived our escape plans from a different perspective. Reinhilde's was based on beauty, mine on language. But if you consider beauty to be the most universal of languages, maybe our strategies weren't so far apart after all. The fact is, we left our homes for universities and international careers. We never returned.

I was the youngest of three by 10 and 13 years—an accident, according to my sisters, but, in my opinion, the jewel in the crown that took a decade of polishing—introverted and quiet, a stutterer with a natural ability to find impromptu synonyms for words starting with p, t, and k. I expected the world to look like my brother-in-law's Desmond Bagley novels, which I devoured well before I should have. A six-week stay with a Tamil family on the outskirts of Colombo made me aware that reality was way more compelling than the plot of novels. I was 17 and hooked on whatever the world had to offer.

Reinhilde began her global journey as an au pair for a wealthy French family, a grand entrée into the world of sophistication, class, and elegance, with walks in Parc Monceau, shopping on Faubourg Saint Honoré, and summer couscous dinners for friends and family at the country home in Pont-Saint-Esprit. She, too, was 17 and just as fascinated by a world where beauty and *art de vivre* could thrive. Theirs was a world without its rougher edges.

These experiences shaped our perspective on life.

Our careers in fashion and business permitted us to travel the world, both professionally and privately. There were years when our combined travel schedules meant spending over 200 days abroad, mostly alone, often staying every night in a different city at a different hotel. We were nomads who got glimpses of cities through the windows of taxis and hotel rooms. Time with people we met and admired was too short. So we compared notes and kept a list of places that we wanted to visit again and people that we wanted to know better.

This book is about these places and their people.

It's a visual chronicle of fragments that share one characteristic: beauty. Never the beauty that is conspicuous or mundane; let's leave that to Instagram and the majority of its influencers, or to the cookie-cutter homes and hotels you find on Airbnb and Booking. It is rarely about the beauty that screams for attention; you'll find that yourself.

It is about the beauty that lurks hidden, that is masked or shy, and that tries its best not to be noticed. It resides in minute details, in juxtapositions, in contrasts, or in combinations. Most often, it appears at second sight. Like the houses on the Garzón town square, squat and unadorned, at the precise moment when their washed-out walls reverberate in the late afternoon sun. Visit the restrooms at the Fondazione Prada, not because you need to go but because Rem Koolhaas turned a purely utilitarian space into the perfect minimalistic piece of art. That's what we think beauty is about.

This book is about time, or rather, timelessness.

What makes true beauty transcend time so effortlessly? A fifth-century BC sculpture of a horse at the Acropolis Museum in Athens is as exquisite as the Kapoors and the Ai Wei Weis that blew us away at the Venice Biennale. Why does the Quadrangle of the Nuns in Uxmal feel every bit as forward-thinking as Oscar Niemeyer's French Communist Party Headquarters in Paris? More than a thousand years apart, they fulfilled mostly the same functions. We cannot explain this through theories; this is not a dissertation. We can only share examples of what we have seen during our travels. It is universal. We found timeless beauty in form, in color, and in light everywhere around the world, without exception. And precisely because of the light, the moment would be fleeting if it weren't for the photograph itself. Talking about contradictions...

As a fashion designer, stylist, and creative director, Reinhilde has a well-trained eye for trends and hype. Don't look for them here. It's not about promoting destinations, restaurants, shops, or events. It's not part of some big-brand marketing campaign. It's a carefully curated selection from a personal database that contains a countless number of photos, taken by Reinhilde, over 30 or more years. You'll see an evolution. We changed over the years. Our teenage sons who joined us on our travels are adults now and travel with their partners. Our dogs—you'll get to know them; they accompany us everywhere—were pups, grew up, and died. Their offspring took their place. In all of this, we find the photographs taken then to be as captivating as the ones taken now.

This book is about the valuable things in life.

We rarely travel because of a destination. We mostly travel because of a house—a distinct place where we can stay and truly discover the area within range of our temporary home. We never would have chosen Joshua Tree as a destination if it weren't for Acido Dorado, the glass house that blends into the desert. We are not talking luxury; the sea wind blows through the wooden slats at Hatch House. The chalets in Borca de Cadore are barely insulated and turn freezing at night in winter. We're talking about uniqueness, authenticity, and creativity. Genuine places designed to meet the fundamental needs of wellbeing at the time they were built. They are getting rare.

Food is one of these basic needs. First on the list at every destination is finding local ingredients, cooking them, and eating at a beautifully set table. Some people take their own pillows when they travel; we take tablecloths, napkins, and silverware. Lobster pasta in Cape Cod, honey and bacon on peanut butter sandwiches in Miami, and Iberico Bellota ham in Extremadura simply deserve to be presented well.

That goes for restaurants, too. They are the places where local families go for their Sunday lunches. They have been around for decades, and they are always packed with regulars. You'll be vetted at the door and served by waiters who take pride in their profession. There will be white tablecloths and restaurant-branded service sets. It will be loud and unforgettable. It's their unrestrained hospitality that made our family feel like locals, too.

Then there's home—the places we always come back to: a 1906 light-filled artist home in Brussels, built for the painter Firmin Baes, and a tiny 1830 farmhouse in a quaint village called Waulsort. We live indoors in one and outdoors in the other. Both were built with a purpose and meticulously restored to keep them very much as they were designed, flaws and all.

Our Brussels home is still a hub for creativity and creation. We filled it over the years with a playful mixture of art, mid-century furniture, books, magazines, photos, and the artworks that our children created when they were young. It is by no means minimalist, but in a very personal way, every object that we have collected seems to find its place organically and begins to radiate inspiration.

Waulsort is where the weather sets our agenda, and the garden commands our activities. In contrast to the dominance of nature that envelops it, this home fits like a tight, warm glove. Only a 90-minute drive from Brussels, the countryside eradicates all traces of the city. You'd expect the two separate worlds to collide, but they don't because Reinhilde designed them as family homes first and foremost, similarly inviting and remarkably connected.

This book is also about happiness.

Each journey is a feast. We have always been passionate about the challenges of new environments, cultures, and customs. Being out of our comfort zone keeps us on our toes. Sometimes we struggle, but we always learn. With more availability and better technology, we can travel for longer periods at a time, working remotely. This makes us more dependent on the local population and economy—more residents than travelers.

We are not nostalgic, but by going through Reinhilde's photo collection and creating coherent and intriguing visual narratives, we relived the excitement of experiences that were once new. Every page illustrates her relentless search for all things unconventional and enchanting. There are photographs that say it all in one single image. And there are compositions that tell a story far stronger than a single image can convey. Light and color always play the leading roles: the washed-out pastels of Salento, the vibrancy of Yucatan, and the earthen somberness of Argyll and Bute in autumn provide the backdrops for her finds.

And for me, there's the extra thrill of going back to creative writing after a lifetime of writing business reports and commercial go-to-market strategies. Don't look for sophisticated essays. I didn't want to write a travel guide, either. You'll find 46 short texts, one vignette for every episode, with a personal exploratory narrative that reflects on contexts that are somehow attached to the extraordinary places, moments, and people encountered over years of traveling. I tried to make them vivid and engaging. Adding a decent dose of humor and a pinch of absurdity threw me back to my days at university. Only the beer was missing.

But most importantly, there's the pride and joy of being able to share our work with an audience around the world, hoping that you'll see the same glimpses of the extraordinary, experience the same sense of harmony, and feel the same emotions as we did when the photos were taken. Let us know; nothing would be more rewarding.

Ours is a very personal view of reality. It is not a representation of the world as it is, or as media platforms impose upon us. Their business model needs banality or polarization to attract views and generate advertising revenue. We don't. It's how we see the world when we travel for longer periods of time with a very selective mindset, dedicating time to photograph and write. You may find this myopic and naive. True perhaps; but why place triteness on a pedestal yet again?

You may think that we are old-fashioned. I disagree entirely. If this is clearly not about fashion, how can it be "old"-fashioned? Our world is speeding up. Business cycles contract everywhere. Industries boom and crash in months. Art and music go viral in days. In no time, the next phenomenon will replace them.

We, too, live in this über-dynamic world.

But, for now, let's take a breather and look forward to enjoying a moment of unveiled beauty together.

HOME | BRUSSELS

INDEX
H
ME

ON IMPERFECTIONS

Someone recently asked us how long we had been together. Reinhilde and I gave different answers. To be specific, the number she came up with was lower than mine. As usual, men like to flaunt their success. And yes, we're still a couple. So it is fair to assume that the discrepancy originates from when our collective history—supposedly—began.

I knew of her before we met. Sheltering from incessant rain and dismayed by the lack of sympathy French drivers were showing for hitchhikers, my travel companions had eulogized her in the deplorable roadside bars where getting dry took longer than getting drunk. Their morale needed lifting. They used adjectives like beautiful, peculiar, and unreachable to describe her. I listened and smiled, amused. My curiosity needed no further bait. My imagination did the rest.

Reality seldom concurs with imagination, at least not mine. When we met a few months later, I searched in vain for the aura of unattainability that they so vividly described. They were right that she was beautiful. Her hair, cut short during her summer holidays and intentionally unkempt, left my friends speechless. They ran out of adjectives. I didn't. To me, she looked well-educated, courteous, and bourgeois. All of which I was trying very hard not to be.

Our common history didn't begin then. She was serious. I was weird; I was in my dog years. Jan—whose father owned a chain of Ford dealerships—and I terrorized the town's campuses and bars. He hobbled, hunch-backed like Catweazl, his duffle coat inside out over his head, as I ran alongside him, barking like a rabid dog and just stopping short of sniffing lampposts.

Reinhilde was a business graduate and conscientiously completed her assignments; I woke up at 2 p.m. and did nothing. She prepared for exams; I read The Hitchhiker's Guide to the Galaxy and watched Dallas on TV. She wanted excellence; anything minimally viable was more than OK with me.

I was anything but her ideal of perfection. There were more cracks in my teapot than unscratched ceramic. And the cracks certainly weren't repaired with gold. They still aren't.

Nevertheless, something clicked between us. It wasn't me who took the initiative; my state of mind wasn't predisposed towards anything even remotely productive at that time. I guess it must have been my older sister's unexpected and flamboyant appearance at my birthday party—she arrived in my father's tan Mercedes coupe donning oversized sunglasses—that must have sparked her interest in me. Soon, friends talked about us as the beauty and the beast. I wore my badge with honor while it lasted. The clouds of bourgeois normalcy had already started to gather on the horizon.

Reinhilde went to Amsterdam and rented a room in a squat full of disillusioned hippies who faced extinction and punks who saw no future. (They all became well-paid managers in the end.) She dyed her hair blue and cut mine. In winter, we skated on the frozen canal in front of the house and got in bed to get warm again. She was in fashion, while I worked in an abattoir.

She shone as an eccentric, and I grew dull. Opposites continue to attract; we're still together. There are cracks in both of our teapots now. Several patches have a delicate shimmer about them.

So what about the number mismatch? Who cares? We've been together a long time. So long that it doesn't matter if she prefers to drop my dog years from her count. We'll probably never get to matching numbers. Not unless she starts barking anytime soon.

Fat chance that this will happen; she's not that imperfect just yet.

D

AFF
ICH
ES
Saïdane Afif
FON
TAI
NES
Saïdane Afif

週

Gabriella Crespi/

Hello I am Erik
Future Present
Congo
Kleyn
BAUHAUS

CHINATI
BOX / 2004
BERNARD BUFFET
CONSTANT
AMERICAN FASHION
HAMPTONS
HUGO CLAUS MAGIER
COBRA
PAUL AUSTER
Glamorama

ANUSJE VAN ALLES

Kleyn

STORIES OF GLOBAL LIVING

GOD IN FRANCE

Some say that France is the country that you drive through as quickly as possible to reach Italy or Spain. They are—mostly—wrong. Some say that France would be the greatest country on earth, if it weren't for the French. They are—mostly—wrong too. But when we talk in German and Dutch about living like a god in France, you, English speakers, are obliged to live in clover.

Considering that God probably knows a thing or two about the quality of life, I prefer France over clover anytime. I'm not a sheep.

Unfortunately, the French have an innate ability to unnecessarily complicate things. The simpler it is, the more complicated they will make it.

Take oysters. Displayed on crushed ice at the "écaillers" of restaurants around the country, they are a mouthwatering must-have. But what should you order? Belons, claires, fine de claires, spéciales, n°1, n°2, 00—does this make sense to you?

Read on if it doesn't. The numbers indicate the size. The higher the number, the smaller the oyster. We're in France, where this makes perfect sense.

"Plates" or "creuses"? "Plate" or flat oysters have a distinctively flat, smooth, and round shell. Belon is the most common one. It has a more intense, nutty taste and a firmer texture. Cupped oysters ("creuses") have a deeper bowl-like shell, holding more meat that tastes sweet and briny. The texture is mostly plump and soft.

"Fines" or "spéciales"? Depending on the amount of meat in the shell, the "creuses" get a qualification of "fines" for the less meaty ones and "spéciales" for the meatier oysters. Only in the Marennes-Oléron area, near La Rochelle, are the oysters finished in ancient clay salt pans called "claire". They are sold as "fines de claire", "spéciales de claire" for the meatier ones, or "pousses de claire" for the top of the line. The Gillardeau was President Mitterrand's preferred "creuse". And mine.

You'll need 6 to 9 oysters per guest as a starter, or 12 to 20 oysters as a main course. The French mostly eat their oysters raw, with a "mignonette" sauce of vinegar, shallots, and pepper, and accompanied by rye bread.

I prefer to eat them with a dash of lemon and some freshly ground pepper. I also generously spread churned fleur de sel butter on the rye bread. Oysters are a feast; they deserve champagne.

God in France, anyone?

18
BFT 77

HUITRES FINES
a EMPORTER
du Producteur au Consommateur
BOULAN Ostreiculteur

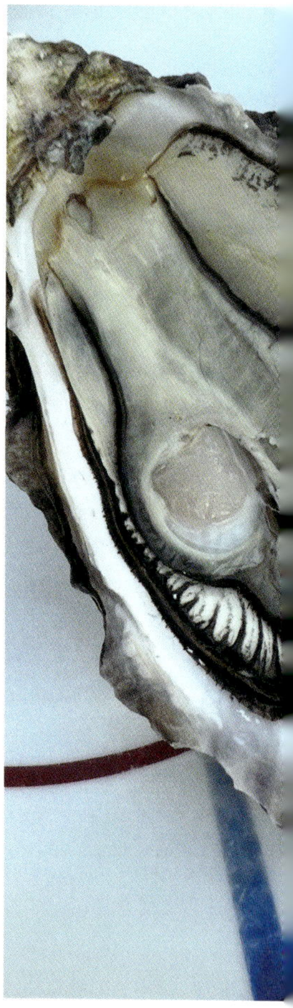

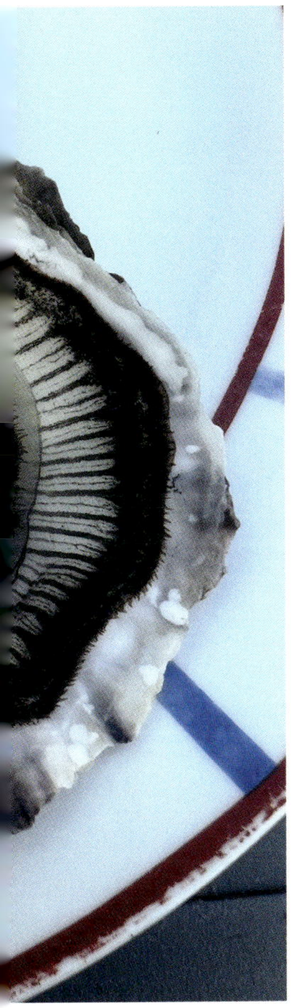

IL MEZZOGIORNO IN BIANCHINA

Cara Sofia,

I dreamt last night.

I'm sure you think so what? Most people dream at night; why shouldn't you? That's true; I know I dream, but I seldom know what about. Imagine waking up passionately in love yet not knowing who with. That's how it feels.

Can you see the fix I'm in?

Curiously, I remember what I dreamed about last night. You were there. Fear not, it wasn't erotic. Perhaps I should be more concerned about the absence of my arousal than about dreams I cannot recall, but let me share this one before it fades out. Francesco Mutti, the graying gentleman with the sideburns and suspenders, was there too. A whiff of passata lingered around him like a chic cologne. His custom-made shirt was stained from a desperate dive to save the last tomato. I'm sure you've seen the commercial.

People were traipsing in and out of an old, elegantly restored house. A table was set in the garden under the pine trees. Everyone was talking at the same time. There was laughter and discussion, gaudy and graceful, like a mundane Galbani spot featuring insalata caprese and red wine.

OMG Sofia, I dream in the clichés that marketers concoct.

I worked the room like a pro. It's an activity I hate, believe me. I'm a lousy networker, an introvert pur sang, who hunkers down into a corner with a glass of wine. But I engaged effortlessly with guests in Italian, French, and English, showcasing wit and eloquence I normally do not possess. There was surprise and awe on people's faces.

I couldn't hear a thing, however. I have no clue what we talked about—soundless lip-synced conversations. Like watching an animated TV debate on mute.

I guess that I am making progress, though. At least I can remember what I saw.

I hope you still park your Bianchina up front with its top down and let the wind play in the pink-striped shades of your house. It's like Cinecitta. Let's go for a swim at Tricase. The port is so small that I can probably make it across.

We'll ask Francesco along, suspenders and all.

Ciao.

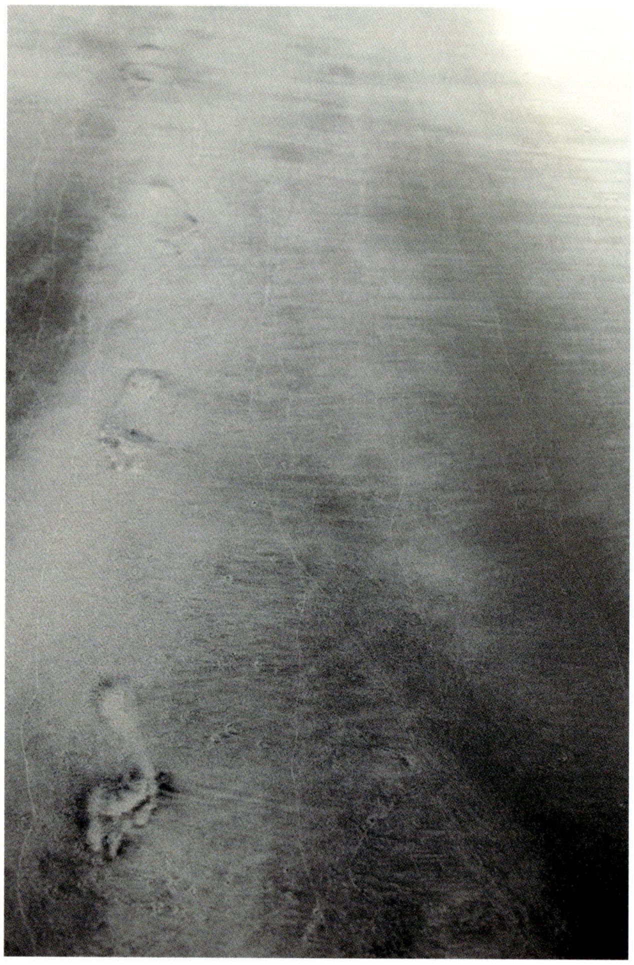

VI35
2790

Portagayola

MÉRIDA MORNINGS

She was already swimming when he woke up.

Their dog had claimed her spot in bed. Sprawled over the crumpled sheets and comfortably asleep, its paw touched his calf ever so lightly.

He marveled at the effortlessness and confidence with which she moved through the water. Even the slightest of gestures seemed to propel her lithe body. Every stroke triggered a bow wave on her forehead. Water resistance didn't slow her. She was on top of it.

It made him jealous.

He'd never been much of a swimmer. His legs always dragged him down. As sturdy as it was, his frame simply couldn't produce propulsion.

So he made coffee instead.

The black, low-cut bathing suit clung to her body when she got out of the pool. Water was dripping from her thighs when she reached for the cup. Tiny puddles reflected the pink of her toenails.

She was just as fascinating as she had been when they first met. His resistance was pointless, too. Her allure still struck him in the midriff.

Mercilessly.

FRESH

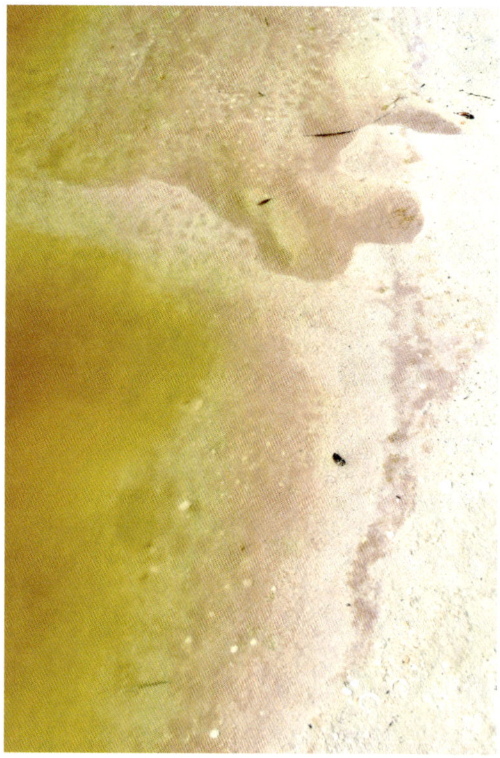

MERIDA
DEL SURESTE

62

ROOFTOPS AND BARFLIES

In the heart of Rome, seagulls shrilled louder than tourists. Just beyond the narrow alley, behind the old chimney of the facing roof, a pair had taken residence. At dawn, they started screeching above the heads of the travelers who were making their way home, the wheels of their trolleys clattering over Rome's unevenly cobbled streets. Two chicks, spotted and thick with down, clambered over the clay roof tiles, vying for the scraps that the adults regurgitated.

Each day, they watched. As the days unfolded, the young birds pushed boundaries, venturing further from their nest, but their cries remained, as loud and frenetic as the discussions they overheard in many of Rome's bars.

Over the weeks, he noticed that the owner of the cafe only smiled at blond, middle-aged women who were a bit plump. Without a care in the world, he sat at a counter that had seen better days, swapping money for minuscule receipts and single-portion packs of salty crisps, stored as prized possessions in a glass cabinet behind him. Like a rock group's tour merchandise, his worn-out navy sweater had the bar menu printed on the back, but the prices had nearly doubled. Around eight, he considered his daily contribution to the nation's finances accomplished and disconnected his credit card terminal. "Non funziona" was the standard response to anyone who showed up with a credit card in hand.

It became their preferred aperitivo spot in Trastevere, always overflowing onto the square with regulars and visitors alike, fashionable or frumpy, young and old, Italians and foreigners, all having cheap beers, spritzes, and crisps, all discussing loudly, all eyeing the few tables and chairs that were never free. On Serie A nights, the football fans gathered and yelled above the Italian commentators, who were blaring from a huge flatscreen in the tiny television room. On these nights, all eyes were on one—a radiant young man wrapped in the carmine and gold of an AS Roma scarf, with an allure as if plucked from a Fendi fashion show.

Their other favorite bar was only a few hundred meters away. At Piazza Farnese, regulars made their way to their favorite tables from decisions negotiated at ministry offices, their tanned ankles flashing from polished loafers—horsebit and all—and their pristine shirts contrasting against cool wool suits. It was easy to be ensnared by the sprezzatura, and dressing with intent became their ritual, a way to lose themselves in spirited debates, intoxicating Negronis, crispy bruschettas, and the weightlessness of life in a city where elegance is hereditary.

Rome radiates energy on a human scale.

First, one chick vanished, then its sibling. The once-vibrant rooftop lay silent. The survival chances of flightless chicks falling from a four-story building are minimal. Stray cats take care of unfinished business. Or so they thought. Until they stumbled upon a fledgling by the riverside, its piercing cry a call for protection. As the adult gulls swooped in defense, they understood the tenacity of life that, for centuries on end, continues to cling to this city.

EJ 976DX

PIVS·IIII·PONT·MAX
PORTAM·IN·HANC·AMPLI
TVDINEM·EXTVLIT
VIAM·FLAMINIAM
STRAVIT·ANNO·III

Vincenzo Monaco
Amedeo Luccichenti
opera completa
HOEPLI

miu miu

VALENTINO
GARAVANI

THE LONELINESS OF A SELF-SERVICE VENDOR

The wheels of the shopping cart wobbled as he pushed on from Pagkrati Square. *Sidewalks aren't made for supermarket carts*, he thought, especially not in Athens. *But what the hell?* He stumbled upon it in Keramikos, took it home, and chained it to the wrought iron fence of the tiny park near his flat. He kept all his assets safe.

When he turned the corner, he found Vasileos Konstantinou strangely empty again. Every Greek with the slightest of common sense had left for the islands, escaping the grueling midday heat of Athens that, like every August, clings to its buildings and rebounds from its battered asphalt, to find cool sea breezes and the neighborhood friends they hadn't spent enough time with during the past year. For those who were obliged to remain in town—bored rental car agents, palace guards, sunburnt tourists—life had slowed down to a succession of freddo cappuccinos. Taxi drivers even used their indicators and gave way to helmetless motorcycle riders, who, only during this month, appeared uncharacteristically non-suicidal.

The sun was directly overhead the Panathenaic Stadium, its pentelic marble benches ruthlessly exposed without a single shadow to disturb its geometrical frugality. Beauty at its most brutal, almost painful to observe. At night, floodlit, it loses its austerity for a shadowy elegance of lines and contrasts preferred by locals who sneak through the concealed hole in the park fence to drink their beers and discuss rising rents and low wages, endorsed by antiquity.

If he desired elegance, he'd admire the discus thrower. But he connects with the woodcutter statue; his serene face defies a body that explodes with anger and frustration. As he reached the park, the wheels of his cart, heavy with a dented styrofoam cooler packed with ice, half-liter bottles of mineral water, and a disintegrating aluminum beach chair, started scratching deep traces into the sand.

He set up his store for the day in the shade of an oak tree. On a piece of cardboard, he wrote, "Water, 1 euro." On another, he wrote "Self Service" and pinned it to the lowest branch of the tree. Sitting idly in his beach chair, he wondered if his smug American punters, devoted to impersonal transactions, would understand the ambiguity.

UNSPOKEN INTERSECTIONS

It was around nine in the morning. The chef was in his whites, smoking by the side entrance, where a succession of nameless vans would soon be dropping off crates of organic vegetables, fresh fish on crushed ice, live oysters, and cases of black-footed free-range chicken. Mornings were quiet on Bank Street—just the pervasive hum of New York traffic and the odd hipster on his fixie bike yelling at a car that hadn't given way.

These solitary moments were his. The kitchen, with its insatiable demands, would soon take over—a brief pause between the chaos of last night's service and the turmoil that awaited: vegetarians asking for risotto without chicken stock, lactose-intolerant patrons complaining about the butter in their béarnaise, and bills to be split unevenly between two couples and three singles.

But his moments weren't entirely solitary.

A man and his dog were walking across the street, as they did every morning. He had a deliberateness to his stride, a contrast to the playful exuberance of his dog, that sniffed every patch of sidewalk and every stray leaf as if it were reading the newspaper, curious to catch up on what had occurred overnight in a world that was more down to earth than theirs.

Nothing had ever been said between them. Two people on different paths, intersecting at the same spot at the same time. Kinship between a chef and a gentleman, both looking for a moment of respite or a touch of predictability.

He did it again. As his dog tugged at its leash, he turned the corner and waved, his hand up in the air for just an instant longer than expected, as if the movement was disassociated from his thoughts, a gesture both noble and maladroit. He responded with a tilt of his cigarette, a silent acknowledgement that had become their routine amidst the urban cacophony that was about to take place.

musco

NEW YORK, 2022

THE PLAZA

EXIT

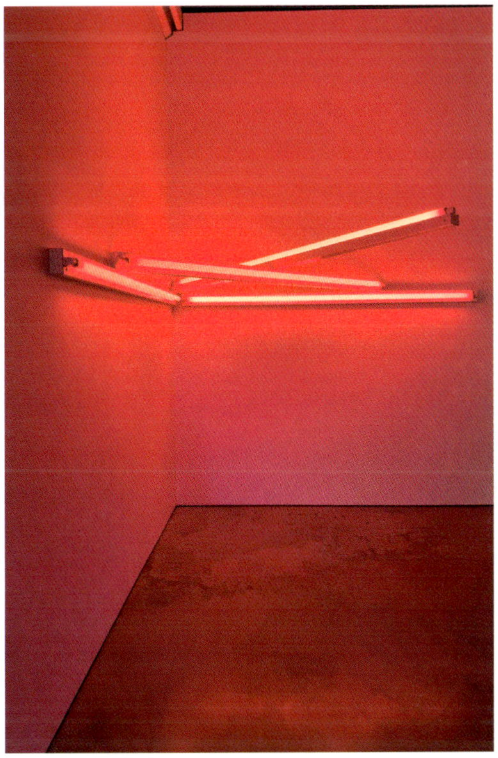

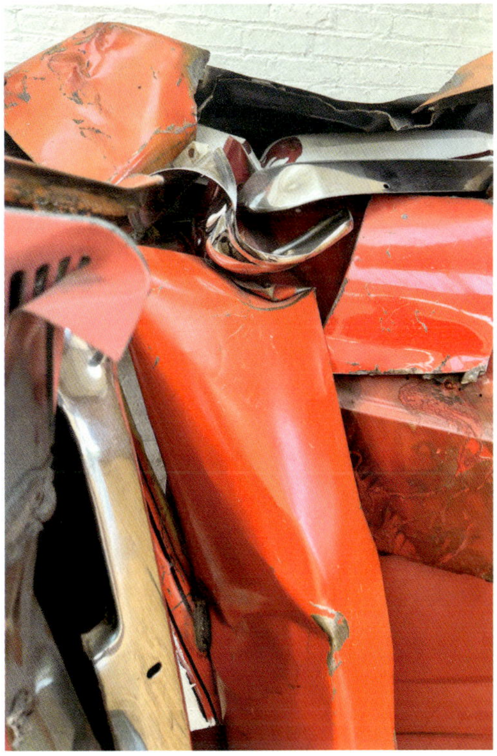

22 IFL 187
BATHROOM MADAM
NEW YORK
C/5

4Y10
Supreme

CLINTON ST
RIGHT LANE
ONLY
ONLY
SIGNAL AHEAD
SPEED LIMIT 35
Jeep

NEW·YOR

BETWEEN RADIANCE AND RESILIENCE

You can tell the city used to indulge in its importance. Its luster still shimmers through the daily bustle on the stately avenues, even amidst buses belching smoke, honking, and accelerating as pedestrians cross the road.

Buenos Aires is unashamed of its greatness, reflected in every facade, polished brass nameplate, and door handle. In the intricate wrought-iron gates of the mansions on San Martin or Alvear. In how elegant men in suits sip Malbec at 5 p.m.

But greatness weighs on a metropolis of 15 million in a country that is, theoretically, bankrupt. Where people queue in front of food stores as soon as they get paid. Because rampant inflation pushes prices up by the time their next paycheck arrives. Where a kiosk on Avenida de Mayo sells more pesos than flowers. Home-delivered if you prefer.

And the porteños?

Those you meet are lively, pragmatic, and beautiful. They tackle yet another political or financial crisis with a shrug and ingenuity. Wondering candidly what on earth brought you to this city that has been dealt such a bad hand.

Why exactly did you come here when you could have gone anywhere you wanted?

So?

Maybe you've been dealt a few bad hands as well. Maybe you were seeking change, leaving the safety of your home in search of disentanglement. In search of beauty that is less obvious, less blatant, less ephemeral, and less superficial.

3169

Suipacha
1400 - 1300
Comuna 1

CHACHINGO
PORRON $
IPA
AMBER
PORTER
PALE ALE
TIRADA $
IPA
IRISH
GOLDEN
AGUA CON O SIN GAS $
GASEOSA $
CAFE $
SPRITZ DE LA CASA
PICSA

HARRODS

WC

FROM CLOONEY TO CATASTROPHE

Marrying a fashion designer has its advantages. You are always well-dressed. That's assuming you are given detailed daily briefings and strict sartorial instructions that you follow to the letter. But when you are given the freedom of choice—even the tiniest leeway, like choosing your socks—things can go disastrously wrong. At least, that's my experience.

I have been married to a fashion designer for a few decades. During all this time, I felt like a star—think of George Clooney, Brad Pitt, or President Macron—having a personal stylist. She decides on the clothes I purchase and she puts my outfit together. When she can't, I just wear the same clothes day after day. That's why we always buy my wardrobe in multiples: three white chinos and five long-sleeved navy polo shirts. Four gray cool-wool pants to go with the fifteen monogrammed shirts acquired over the years. It's called risk management.

It is solely because of her that I get compliments from time to time. The ones that matter are praise from other well-dressed people. The admiration of women wearing lycra tights and oversized sweatshirts at gallery openings gives me little satisfaction. Nor do I sympathize with the envy of men wearing cognac-colored brogues that are laced in blue. But it's hard to suppress a blush when, out of the corner of my eye, I notice a sophisticated woman giving me a second look. And when she's dressed fashionably, I get palpitations and feel uncomfortably lightheaded but radiant.

I faithfully follow my wife's instructions. But every so often, they are incomplete. I think that, cruelly, she puts me to the test sometimes, leaving me with the agonizing task of making up for whatever she left out. I break out in a cold sweat. I desperately try to come up with something that matches my uncompleted attire for the day. Or, even more challenging, something that she will approve of. I apply strategy and intellect, but neither of them provides any backing. In the end, I pick whatever lies on top.

Regardless of the methodology I use, my choices are seldom right.

On a good day, I arrive downstairs fully dressed and reasonably cocky, prepared to meet *the look*. She scorns me within seconds. She puts her hand in front of her eyes and eradicates the eyesore that is me.

"How on earth can you wear those socks with these shoes on a beautiful day like this?"

I still feel that *because they were on top* is a perfectly logical answer, but wisely hold my tongue. Confused, helpless, and heavy-footed, I go back upstairs, put on the socks that, for reasons way beyond my comprehension, go well with the shoes and the weather, and return for breakfast and a rigorous second inspection. She'll straighten my shirt, pull at my lapels, and tell me to do the buttons on my jacket, but only the second from the top. I leave home upright and confident.

Bad days are far worse.

Bad days are when I'm not scorned in the morning. When she feels that there is no need to eradicate the eyesore. Bad days are when I walk out straight and self-assured into a busy day, convinced that I've got it all under control, and proud of my vestiary nous.

But when I return home, she looks at me from top to bottom, shakes her head, and puts her hand in front of her eyes.

"Have you been out all day looking *like that*?"

Cartier

CAMPARI

Dominique Ropion
OF A LADY

STOCKMAN

RICHARD MILLE

U-24

Alexander
McQUEEN

Pirelli HangarBicocca

malo

FALCONERI
SUPERIOR CASHMERE

5

BEYOND RESISTANCE

José Ignacio appealed to us the moment we arrived. Totally irrational. It's the heart that overrules the brain, very much like love at first sight. Sure, it's a unique village, largely unspoilt, and surrounded by an ocean that's simply overwhelming. And yes, it's easy to be charmed by beautiful people who are on holidays, toddlers in tow. But there is more than just the allure of easy and carefree lives. This place creates a stronger and more compelling bond—an interconnectedness that beckons beyond resistance.

I

Sofia's thoughts were as murky as the water of the Rio de la Plata that she had crossed on the Buquebus ferry to Uruguay.

As she was about to depart, her father told her the story of how God created Argentina. Countries complained to God that he gave Argentina too much of everything: too much beauty, too many resources, too good a climate, you name it... Hearing their laments, God admitted his fault. To make up for his misstep, he populated the country with Argentinians. They squandered it all.

She wasn't as sarcastic as him just yet. She felt that Argentinians were the country's best asset. But God hadn't exactly bestowed on them the capability to elect inspired leaders.

The cost of renting a cramped apartment in Buenos Aires has skyrocketed over the past 12 months. She couldn't afford it anymore, and if she didn't make enough money this summer, she'd have to move back home with her family. And spending all day dodging her dad's snarky remarks? Not something she was thrilled about.

As the low light rose and pierced through the clouds on the horizon, the ocean was liquid silver. The waves crashing on the empty beach were hissing as they receded. Birds and crabs had devoured half of the fish that did not make it back to the water in time last night. In the distance, a lone angler was trying his luck to capture something alive. *Why does an angler dress in white tennis clothes,* she thought.

She walked back towards the empty beach restaurant. It had been built with a nonchalance that didn't match the prices on its menu. Long tables were placed under beach tents that were covered with tarp and reed. The chairs were weathered and rusting. Last night's staff left the fridges unlocked. But nobody minded.

Getting drunk at six in the morning wasn't on her mind. But it sure as hell might have helped

||

Gabriel crammed the old Volkswagen Kombi with plywood shelves, barely leaving enough space for him to squeeze in. He loaded up boxes of second-hand novels from Corrientes Avenue bookstores before setting off northward, leaving Buenos Aires behind, stretched out in all its squalor and splendor. With a few hundred books to sell, he hoped to stay financially afloat for a summer along Uruguay's country roads, rural towns, and beach villages.

He wondered how the customs officials would react upon seeing him enter the country in a rolling bookshop that was barely disguised as a camper van. The outcome was as simple as it was unexpected: they didn't.

He hadn't noticed that he had parked his van in front of a bakery, but the sounds of chefs banging their utensils with what appeared to be utter frustration with stainless steel made him painfully aware that he had barely slept, that his books were still short of punters, and that his wallet was desperately short of pesos.

His bookshop on wheels had made little impact in the sleepy towns where he had spent the nights since he entered Uruguay. They all looked the same with their shaded town squares, squat whitewashed houses, and inquisitive neighborhood dogs that sniffed his pants and pissed against the dented chrome bumpers of his Kombi.

The air was heavy with brine when he set up shop for the day. He squeezed his mattress and clothes into the passenger seat of the cab, making space for people to crouch inside the van and read the spines of the books he had arranged alphabetically by the author's name.

A slim young waitress in a blue jean apron walked to and from the kitchen, serving pastries and coffee on the bakery's terrace. She moved with the ease and poise of a ballet dancer. Her fringe was too short, bobbing up and down with every stride. She appeared to have stepped right out of an Almodovar film.

That's when he decided to stay.

III

Frances blamed the splitting headache she woke up with on the last bottle of Tannat. They'd spent the afternoon at her parents' friends' polo grounds a few miles inland, watching practice games and catching up on the gossip she'd missed while in Buenos Aires. The grounds were immaculately kept, and food and drinks were served in the new hospitality chalet, an outbuilding near the stables with a professional kitchen and staff, that could easily accommodate a party of a hundred or more.

The conversation among her parents' friends—the Argentine business elite who had relocated their family and assets to Uruguay's clean ocean beaches, verdant hinterland, and discreet banks—quickly turned to politics, the inherent distrust of their leaders, and their inability to manage a country rich in everything except political vision. Working in BA, she saw how people were forced to devise the most inventive strategies to deal with the unending crisis called Argentina. She admired their unwavering resolve.

With the surf forecast predicting a one- to three-meter swell, she decided not to nurse but to fight her headache on the waves. The muscles on her tanned thighs stood out against her springsuit. She'd spent a lot of time in the gym recently, and she hoped that the strength she'd gained in her legs would allow her to pop up faster and ride out more easily.

When she left, the house was quiet. She walked past the pool and down the steps onto the beach, the dark blue funboard with the bright yellow bottom tucked beneath her arm. She cherished the early mornings, when all she could hear were the roar of the ocean and the squawking of the birds. The water was colder than she had anticipated as she walked into the white froth and dove beneath the wave without hesitation, only to resurface trembling, alive, energized, and ready to ride.

She parked the old Land Rover on the corner of the bakery that made the best viennoiserie in town and ordered a cappuccino and croissant. The waitress had to be new; she hadn't noticed her here before. She walked as if on stage, determined and elegant. When she brought her breakfast, they both spotted the vintage Volkswagen Kombi parked on the curb. It was blueish-green with a white stripe and dented chrome bumpers. Just then, a young man stepped out of the van holding a stack of books.

He turned around and looked at them.

parador
la huella
faro josé ignacio

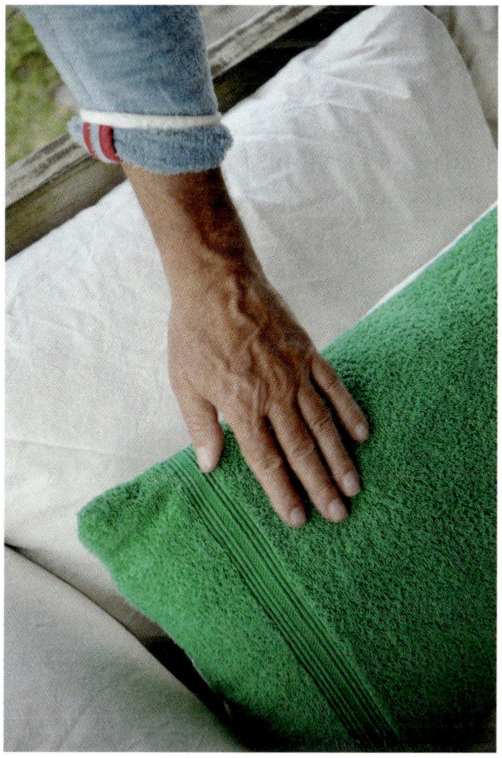

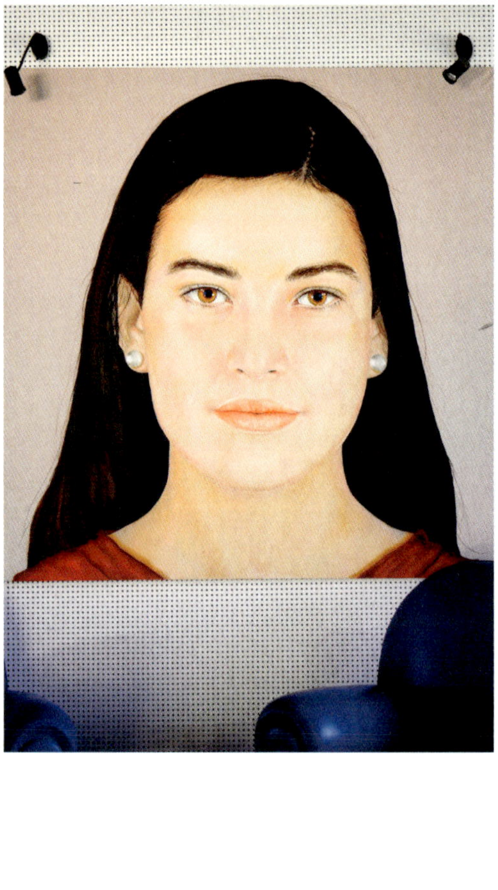

libros

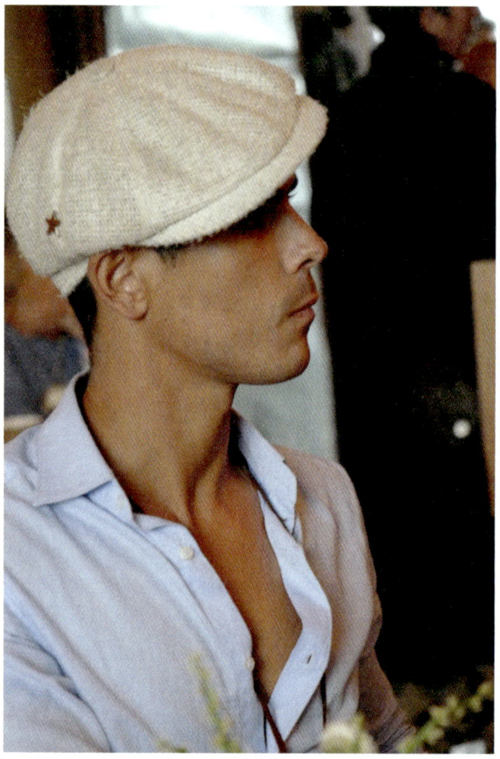

CAPE COD LOBSTER PASTA

Cape Cod lobsters are amazing and about as reasonably priced as you can find anywhere in the world. You'd be a fool to pass up a plate of freshly made lobster pasta for lunch, especially if you can eat it on the deck of a 1960s modernist beach cottage while gazing out at the bay and surrounded by nothing but nature.

Here is how we prepare lobster pasta for 4 people:

Ingredients

2 live, hard-shell, 1.5-pound lobsters, bought just before you start cooking

2 lb. of cherry tomatoes

For the court-bouillon: 1 stalk of chopped celery, 1 big chopped carrot, 1 chopped white onion, a bouquet garni, black pepper, and chili flakes

Half a white onion, diced

6 cloves of garlic, chopped

Olive oil

Salted butter

Fleur de sel, ground pepper, and chili flakes to taste

Spaghetti or linguine

Recipe

For the sauce, roast the cherry tomatoes in the oven until their skins start to char. Switch off the oven and let them cool down and caramelize in the heat.

Bring the court-bouillon to a boil in a pot large enough to hold both lobsters.

Remove the rubber bands from the claws just before you put the lobsters head first into the boiling water. Make sure the water boils again before you put the second lobster in. Cook each lobster for eight to twelve minutes, depending on their size. Take them out and let them cool at room temperature.

Remove the skins and keep the peeled tomato aside.

Remove the lobster meat from the claws and the tail. Tear the meat into large chunks. Leave the claws as one piece, if you prefer. Remove the small legs and use them as finger food.

Bring salted water to a boil for spaghetti or linguine. Cook al dente.

Heat enough virgin olive oil in a large, heavy skillet to sauté the onion and the garlic over medium heat.

Add the tomato when the onions are translucent. Season with fleur de sel, pepper, and chili.

Use extra olive oil to finish the sauce and add the pasta. Toss thoroughly to ensure the sauce is evenly distributed.

Melt the butter and briefly sauté the lobster meat.

Arrange the pasta on a warm serving plate with the lobster meat displayed on top.

This dish deserves to be paired with a rich, oaked Chardonnay, a refreshing Albarino, or, if you're turning this meal into a special treat, go for champagne, preferably a 100% chardonnay blanc de blancs.

after construction, on what was then a treeless moor

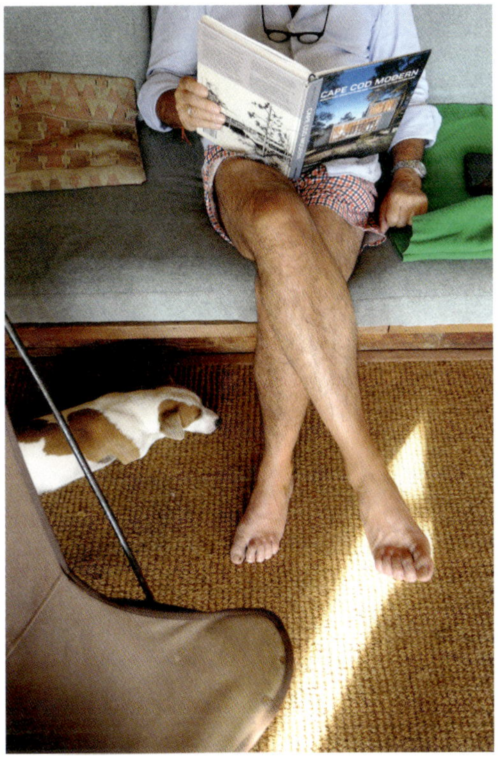
CAPE COD MODERN

SHOOTING INTO THE CORNER
(ANISH KAPOOR)

Venice floats on light, she softly said,
Laguna blue. Just like her eyes reflect
The neon hue of hoses on the market floor, still wet
From cleaning. The smell of fish. And raucous voices take effect

And ricochet. She gazed at drains disgorging water onto streets and squares.
It's the shimmering invasion of high tide, abruptly ankle-deep,
That makes the passerelles appear—a measured pace that keeps
Us moving antlike sets off a restlessness we are unable to unshare.

We have the time, she murmurs, but I doubt
If in this slow-mo floating world where non-stop scenes unfold
There will be time enough to hold our breath and just observe

Man's deepest black and mattest white that's subtly bulging out
From pregnant walls, or allow ourselves to be cajoled
Into life, and birth and bloodshed and let the carnage hit raw nerve.

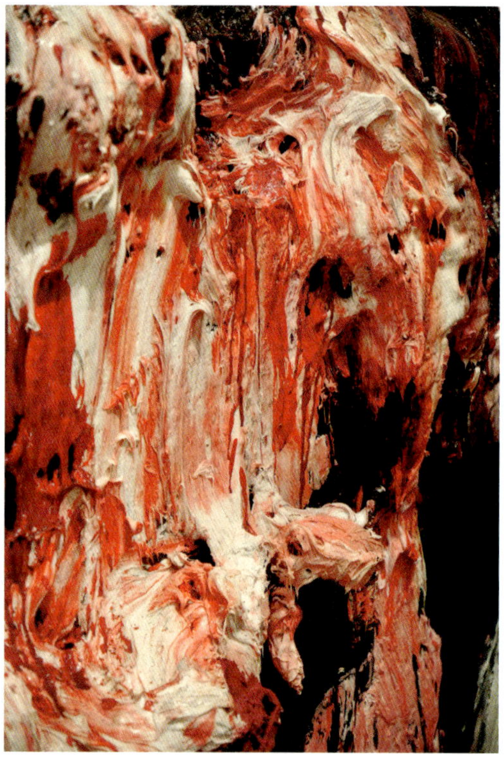

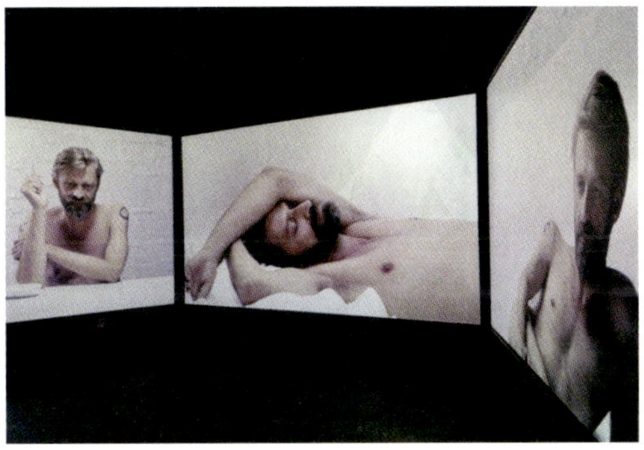

THE BALAYAGE OF GODDESSES

"Hi there. Can I help you?" She was standing tall at her maitre d' station at the entrance of the restaurant and immediately regretted what she had said. She had been looking at the reservation list as they walked in.

My god, they're European! They'll probably answer my question, she thought. *They'll be eloquent and reply in perfect sentences, with subjects, verbs, the lot. Posh accents*!

She could easily tell Europeans apart from her New York clientele by their faux low-key attitude. Underneath it lies a sense of superiority under the guise of courteousness or wit. *To them, we'll all be philistines forever. They are lousy tippers too*, she knew. On holidays, they are shabby chic and wear their battered, monogrammed shirts over designer shorts. *Elegant penny-pinchers—that's what they are.*

"Good afternoon," the man replied, "you certainly can. Do you still have a table for three, please?" *There you go—just what I thought.* Two colleagues joined her at the stand, their faces smug. They had been pushing 140-dollar pitchers of sangria to a group of guests. With obvious success. Big tips were in the making. Just what she needed now.

Hello, anybody home? he thought, but he held his tongue. She was perched over the maitre d' station. He could only see her expertly styled hair and bare shoulders, which were muscular and tanned. *They only seem to advertise this job to gym bunnies with a balayage and the posture to stand behind a lectern.*

The place looked like a Parisian brasserie with the vibe of a bagno in Forte dei Marmi, except for the staff, the menu, and the prices. It was buzzing. A German family and their American hosts were having a ball inside. Two other maitre d's, equally toned, tanned, and hair-styled, had joined her at the stand: three untouchable goddesses wielding sovereignty with blatant delectation. They laughed—the empire was under their thumb.

He wished that he could peer through people just the way she did. *She seems to strip my desire of even the slightest panache or innuendo*, he reminded himself. *Don't even try!*

"Sorry, for three?" he asked again, louder now, breaking an awkward silence. Her focus changed.

He hoped his beer would be as cold as the look she gave him.

MASHOMAC
CLEARANCE 10' 9"
PASSENGER CABIN

ICCOZZI'S
ROPANE CO.
MMER HOURS
GMC

Bliss

MARNI

ABA
ABA
ABA

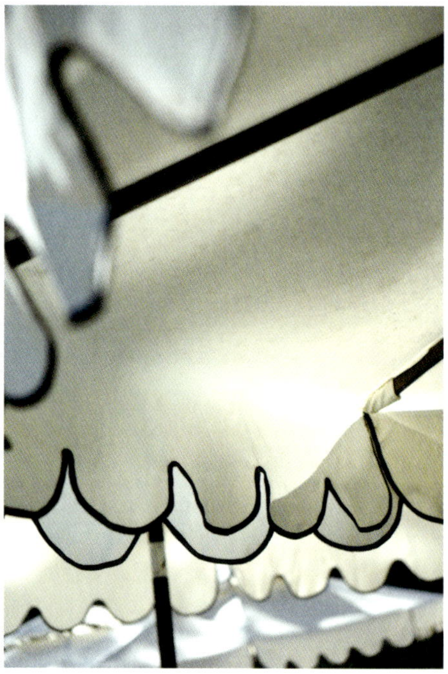

NOT FOR THE UNINITIATED

Don't get me wrong. I'm not trying to settle a score with Georges. Despite the man's relentless ramblings, his annoying display of flamboyant über-French savoir-vivre, and the torrent of condescending remarks he would throw around whenever he was drunk—a condition he often preferred to sobriety, especially when induced by champagne, which was the coveted source of his intemperance—I genuinely liked him. I guess, after all these years, I still miss him.

Paris plays tricks on foreigners that hang around for too long: they trade their origins for a larger-than-life portrayal of parisianism. They overachieve with faultless French and become inimitably arrogant. Georges had nurtured both competencies to perfection and propagated them generously.

I met with Georges in Paris every couple of months for years on end. Mme. Mairesse, an aging accountant equipped with an armor that left her unscathed from even the most determined attack of charm or humor, commandeered the colorless antechamber of the once luxurious but now threadbare apartment that served as his office. The drawing room was rearranged to fit two stately curved desks installed at an angle in opposite corners against a backdrop of faded burgundy flock wallpaper. Despite the unflattering neon lights, Georges' demeanor, his de Fursac suits, and Hermès ties stood out as the only perceptible touches of elegance.

While the day's schedule was routinely identical, the result was mostly unpredictable. Office meetings were kept to the fundamentals of everyday banter, the latest gossip, and amicable horseplay. They were short. Business only started after the first bottle of champagne at one of Gerorges' favorite restaurants.

When things went well, the day ended with a short wrap-up meeting at the office. I'd return to Brussels with a handful of contracts. On other occasions, Mme. Mairesse dropped off my briefcase at the restaurant around six p.m. We'd still be arguing incoherently while the maître d'hôtel stood by, armed with another bottle of champagne. The train ride home was a blur. We had discussed our business, our families, and the global state of affairs. We shook hands on huge deals that were honored unreservedly.

The market changed. Georges didn't. Our business ended abruptly. I never heard of him again.

But he still emerges whenever Paris radiates in all of its generosity, splendor, and gallantry. When its brashness slaps you in the face, utterly unprepared. It's not a city for the uninitiated. But I have had an expert teacher.

MATIÈRE NOIRE
LE JOUR SE LÈVE
SPELL ON YOU

CHANEL

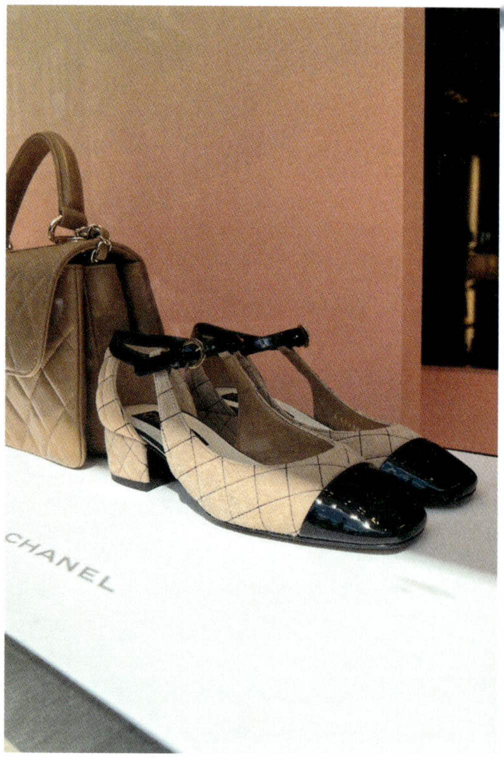
CHANEL

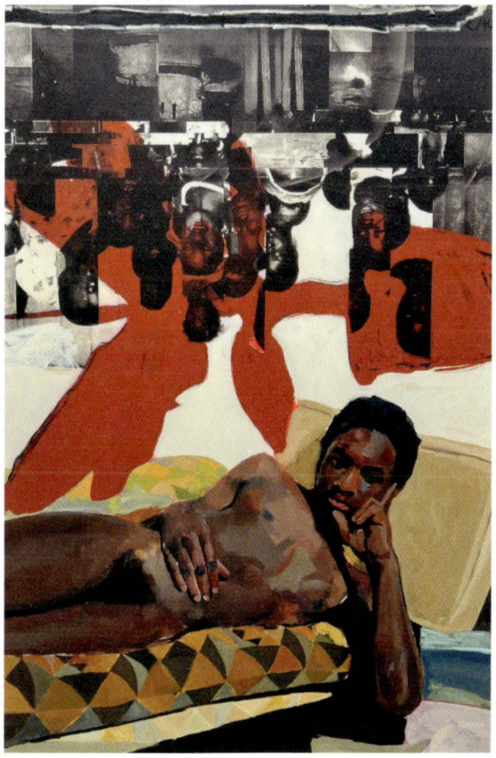

LES EXCLUSIFS
DE CHANEL

LES NUMÉROS DE CHANEL
CHANEL

Magasin Louis Vuitton
2 Pl. Vendôme Paris
Vitrine 5, sacs Twist
Du 13 mai au 7 juin 2022

GERT VOORDJANS

BON

EXTREMADURA DELUSIONS

Sometimes you just get things wrong. A thought that is stuck in your mind nurtures itself over time, feeding on your impressions, misconceptions, and expectations. Never challenged, it progressively grows larger than life. Your life. Not the lives of anyone else.

Delusion is often a solitary experience.

It's a bummer when you find out that you had it wrong all this time. That you held onto your misconstrued conceptions of reality only to find out that the real thing is—how shall I say—a lot more mundane.

I have always thought—romantically, or at least naively—that life would be extremely hard in Extremadura. Why else would you give this Spanish province such a name, after all? That the climate would be harsh and the soil arid. That people lived in hardship, working small plots of parched land with modest harvests.

But the "Land beyond the Douro River" is where Emperor Charles V decided to spend the last 2 years of his life, attending daily mass from his bedroom window because he was suffering from gout and unable to walk. Ruling over an empire where the sun never set, he wasn't exactly short of places to stay. And you don't get gout from malnutrition.

Extremadura is where the red-coated Retinta cattle graze and where Spanish tobacco is grown and dried. It's the land of the Dehesa, the Spanish savannah, with its luminescent flowing grasses and solitary oak trees, among which black Iberico pigs wander and forage for acorns. It's the region that produces the world's best ham, the Iberico Bellota, aged for 36 months to achieve the sweetest possible meat, served at the tapa bars in Cáceres, Badajoz, and Mérida.

As I take another sip of the burgundy-colored Habla No. 7, I realize that getting things wrong from time to time can't be all that bad. And I muse about the right words to describe the feast I once considered suffering.

8 PIEZA
Y MUS
PECHUGA
ECHUGA

MARQUESA
ENCICLOPEDIA CULINARIA
Cocina Gallega

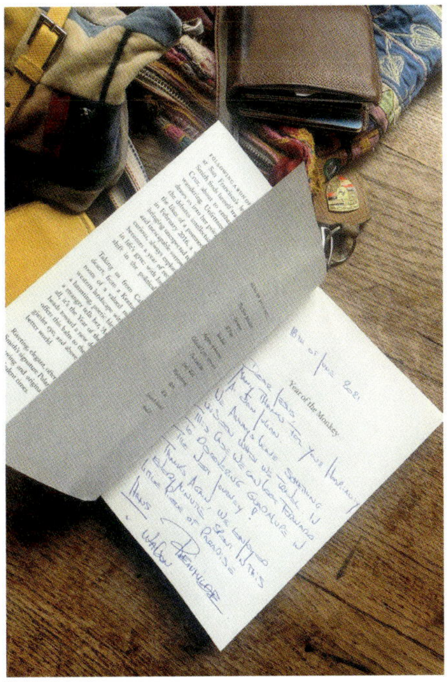

ON TRAVELING

I don't have many regrets. An opportunity arises that requires an immediate decision. You don't give it much thought; you decide, and it ends up amiss. *C'est la vie.*

But there is one decision that I have regretted all my life. It was in the mid-nineties in New York, on a layover from New Zealand. When you are flying to the US from New Zealand, you arrive before you depart. Remember Bill Murray in the Groundhog Day movie? That's how you feel.

Flying was great then; you dressed up for it. British Airways had onboard chefs in Business and First, crew members that remembered your name, and pilots that occasionally invited you to the cockpit for a chat. I was a BA Gold member for life—life at BA is limited to thirty-six months, I learned later—which meant checking in from a dedicated terminal at JFK. Airlines kept their distinguished passengers as far removed from the crowds as possible.

That's where I made my mistake. As I checked in, I was offered a seat on the Concorde to London for an extra 700 dollars. I refused. Even though a standard return cost 12,000 dollars at the time, I found 700 dollars too much. I was an idiot.

I first heard it on the radio in my car on July 25th, 2000. A piece of titanium on the runway ruptured a tire on Concorde flight AF4590. Tyre fragments punctured its fuel tanks and set its engines on fire. The plane crashed minutes later, and no Concorde ever flew again. I missed my chance to fly supersonic.

A few months later, I took my first Ryanair flight to Glasgow and didn't check the airport code. It landed at Prestwick, an hour's taxi drive from Glasgow. On board, I paid for coffee, dealt with the crew peddling scratchcards, and got my eardrums busted by clarions blaring an on-time arrival. The age of no-frills flying had arrived with a bang.

I spent my 1.5 million miles flying the family in first class to Mexico on Swissair for a holiday. We all dressed up for the flight. A week after our return, Swissair went bankrupt and restarted low-cost as 'Swiss'. I wonder if they shortened the name to save on paint.

WHY MALES ARE USELESS

It's only the human species that seems to have found a legitimate use for males, he thought, assuming at least that occupations such as resurrecting imperial ambitions, orchestrating geopolitics, and morphing yourself into a caricature of a reputable presidential candidate can be considered virtuous.

Fortunately, he only entertains incoherent imaginations like these when he is cooking.

He was breading the orecchie di elefante that he would serve for lunch. Egg was dripping from his fingers. He had butterflied the veal cutlets and flattened them with a mallet until they were just the right thickness to be fried in the clarified butter. He'd use an indecent amount of it; the huge chops would float until the breadcrumbs turned golden and crispy.

Elephant ears are the best use ever for a waste product. What else can you do with male calves? Cows provide milk. Bulls produce manure. Treat yourself to a dry-aged ribeye from a fattened Hosteiner or Simmental cow, but leave the bull beef to McDonald's. It's game over once testosterone comes into play.

He felt relieved that there was at least one female presidential candidate.

He rode horses when he was young. He remembers a grainy photo in the newspaper of him as a tiny teen—he had his growth spurt only at seventeen—on a towering horse soaring over a square oxer with inches to spare. For the first time ever, his classmates were in awe. The horse was called Volcano. It was gelded at two years of age but continued to behave like a stallion; a poser, totally oblivious to the fact that he had lost his orbs years ago. There goes our option of castrating our most belligerent leaders, he concluded; they all want to add to their crown jewels, not lose them.

You may not have it made if you are born a male calf on a dairy farm, but can you imagine what existence looks like when you hatch as a male chick on an egg farm? Life's a lottery.

He removed the veal chops from the pan and dried them on paper towels. He rubbed the edge of his knife over the crust and grinned as he heard it crackle. His largest plates barely held the meat. He arranged some rucola salad and halved cherry tomatoes on top, with a wedge of lemon on the side.

Handke
TESORI D'ITALIA

31

A MODERNIST MOUNTAIN UTOPIA

Enrico Mattei had a knack for making enemies. Powerful enemies. His plane never landed at Linate Airport that October night in 1962. It disintegrated. Lowering the landing gear set off a bomb, killing one of Italy's most controversial figures of the 20th century. The ENI CEO singlehandedly dragged post-war Italy into modernism, broke up the US oil monopoly, and contributed to Algerian independence.

And he had the money to create Villaggio ENI in Borca di Cadore.

From every chalet, you see the pink splendor of Monte Pelmo in the early morning sun, with snow-covered pine trees framing it. The concrete floor, painted a bright yellow, contrasts with the red and blue walls of the living room and kitchen. Through the expansive south-facing windows of these minimalist structures, the Dolomite scenery jumps at you in all its intrusive grandeur. This landmark's eloquence baffles you as you make coffee in the intimate kitchen with the yellow Formica worktop. Ageless design at its best.

Mattei was a paternal industrialist. He imagined Villaggio as a modernist utopia in the format of a company vacation town. Breaking down social barriers and uniting the families of both ENI laborers and executives allowed him to infuse his employees with a company-shaped, futuristic Italian lifestyle that thrived on Agip fuel and Fiat 500's. To succeed with this experiment—which was as commercial as it was social—he needed a visionary architect: Edoardo Gellner.

While the few chalets that have been appropriately restored are still strikingly modern, they must have been disturbing in 1956. Gellner's innovative total design concept—comprising urban planning, architecture, interiors, furniture, and accessories—eliminated the traditional aesthetics, dictated by social class, to create a strange new world. Everyone interacted spontaneously at their temporary homes, the church, the bar, or the tabacchi—architecture with social and political objectives.

Most chalets have now been sold as holiday homes. Their windows have laced curtains. The colors have faded, authentic interiors have been ripped out in favor of Ikea furniture, and trees have been cut at random. The tabacchi and the bar have closed. The egalitarianism of mass-produced commodities beat design excellence.

Only the church remains—commanding, intact, but equally abandoned. Another failed utopia.

40

WinteRace
CD 194YS

VOGUE
GLOBAL

NOTHING BUT THE SUN

The town belongs to the obstinate; 250 or so of them are left. Their world went away, but they held on for dear life. Maybe they are waiting for the blunder of resurrection to come about. It makes sense; after all, it was blunders that caused the place's emergence as well as its near extermination.

Is it time that's spurring you? Or is it desire?

Time reigns in Bombay Beach. The past, laden with the sulfurous smell of decay, clings to the broken beauty of whatever has been left behind: once-hip Airstreams, wheelless car wrecks, and bleached 1950s billboards that advertise attractions long vanished. Among the relics of prosperity, a man sits in front of his trailer on an oversized, worn-out sofa—his dog, his cigarettes, and a half pint of Jim Beam by his side. Nothing happens but the sun. Only the chugging noise of Union Pacific freight trains breaks through the appeal of the sluggish and monotonous present here. And the future? What future?

Desire is just a drive away. An hour and a half up the road, at most.

Palm Springs conceals more than just the heat beneath its polished facade. The sleek mid-century homes stand like elusive icons of a sun-soaked elite that sought solace in the desert—mementos of modernity and ambition. Landscape designers transformed the arid desert soil behind their walls into manicured lawns, kidney-shaped swimming pools, and outdoor kitchens. It's where the smart and inconspicuous meet for poolside cocktails and canapes against the rugged backdrop of the San Jacinto Mountains. Design and sophistication nurture a peculiar sense of community in this town, in response to the brutality of nature.

There is nothing in the desert, and no man needs nothing.

Gone are the speedboat trailers on driveways. It's just the SUVs now. The Salton Sea is disappearing. Decomposing jetties stand on dry ground. Beach clubs have disappeared, except for vintage postcards. Only the Ski Inn bar, plastered with dollar bills, remains as a distant reminder of Palm Springs' lakeside playgrounds.

There is so much desire, but is there enough time?

2 ND
SKI INN
SKI INN
World Famous
Lowest BarInTheWesternHemisphere
STOP
COCKTAILS

SILVER

STOP

POSTED
NO TRESPASSING
KEEP OUT

BUMS TO BERLITZ

So this is how mice see the world, he observed, his head sore from lying on the concrete floor of the railway station. He wasn't sure if his idea originated from the bong he smoked with the soldiers on the train from Malaga or from the beers he shared with them. But the black leather boot, only inches from his face and polished to such shining perfection that it reflected the steel skeleton of the platform's canopy, did not look like an illusion.

Confirmation came when the second boot nudged him unkindly in the ribs. Behind a pair of creased khaki pants, a man pulled a wooden luggage cart, its red and yellow paint chipped. A glob of orange crowned a heap of drab suitcases. *There goes my backpack*, was the second accurate assessment he had made since waking up a total wreck.

Without a single comprehensible word, two Guardia Civil officers marched him into a windowless room with a wooden table and two chairs. His decision to spend a few weeks backpacking through Spain—to rebound from expertly flunking all of his June exams—was losing its appeal. Their tricornio hats looked ostentatious and ludicrous. The HK MP5 submachine guns that hung loosely from their sides did not. The officers' lack of laughter was contagious.

It was five past seven on a warm July morning in Granada when the door slammed shut behind him.

His big orange backpack soon caught up with him. It was emptied out on the table while he stripped. They found no drugs. His rectum, too, was officially declared unpacked. In fluent English, the sergeant wished him an enjoyable stay in Spain. *Somehow, looking up my backside stimulates the language center of the Spanish brains*, he concluded. *Maybe I should sell my bum to Berlitz.*

Granada's railway station hasn't changed in all these years. It looks as outdated against the futuristic Talgo trains as a tricornio hat by a machine gun. The concrete on the platform has been polished. It looks softer now. There are no more soldiers carrying cheap Moroccan weed from Ceuta. Big orange backpacks have been traded in for hard-shell trolleys.

He spent the night at a luxury estate in the hills. There was a private chef, a manicured herb garden, and a stud of Andalucian purebreds. The owner made a fortune growing legal cannabis. Everyone spoke English.

He didn't even have to strip.

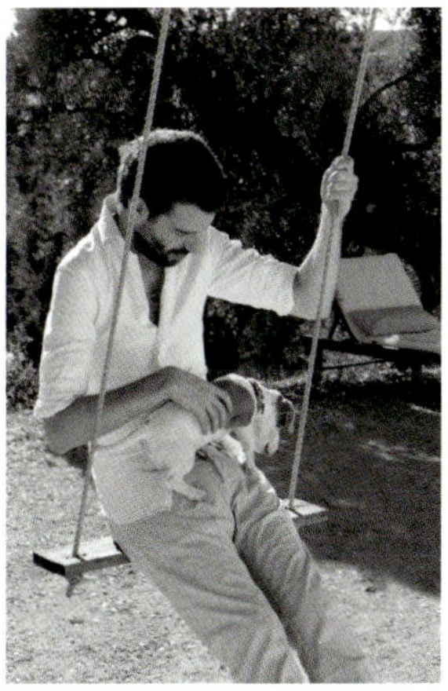

POSTERS

FOXED IN PRENZLBERG

My first memory is hunger and the smell of bratwurst. The smell of bratwurst went away. Hunger has always hung around me like the toxic friend you were warned to avoid.

I feared that the noise in my belly would scare the pigeon away, but it just sat there. Its neck snapped with a single flick of my head, exactly as I had been taught. I strutted along Senefelderstraße with the pigeon in my mouth, its head dangling lifelessly. Some people looked at me as if foxes had no right to use the sidewalk. Others never stopped talking as I walked by.

"I told him I wanted to kick him in the nuts," one girl says. She swaps the paper cup of cappuccino for her right hand, then blows on the fingers of her left hand. She didn't even notice me.

"Who? Your boss?" the other one asks before she takes a sip of her oat milk latte. There is a red smear of lipstick on the cup now.

"No, my coach."

"Why do you want to kick your coach in the nuts?"

"Not my coach, my boss." Her cup swapped hands again.

"So, what did he say?"

"My coach? That my mental health matters."

We listened to the heavy footsteps of the park keepers' boots on the wooden floor of the shed above us. The den was cramped for five kits and a vixen. We cuddled up to our mother. It was comfy. The strongest ones at the front got her licks and her milk. I was at the rear. I sucked the hind tit. There was never enough milk for me.

"It's been great to see you," he said, "after all these years. We really shouldn't have waited that long. Let's catch up again soon." He is on the phone, and he is in a hurry, taking big, important strides. He is lying. I can tell. Cunning is my second nature.

"I'm not sure if March will work for me. I promised Leonie we'd go to Italy. Eat pizza and have espresso. Get drunk on spritzes. You know how it goes. But I'll let you know."

He listened in silence.

“Yes, let’s do that.” He replied.

“I’ll call you back. Say hello to your wife.” He took the phone from his ear, looked at the screen, and noticed that they’d been talking for four minutes and seventeen seconds.

“Mara, yes, of course. Do say hello to her, will you? Tschüß.”

I always eat the innards first. They are juicy and warm. The pigeon was skinny; there wasn’t much meat on its bones. I ate it anyway, bones and all. I don’t eat the wings; I find it hard to digest the feathers. I leave the wings for the crows.

I wonder if birds eating birds is called cannibalism.

It was easy to fall asleep on a full stomach. I dreamed that someone left a box of Standard Serious Pizza in the alley. Nduja wafted through the soggy cardboard—the easiest meal in months. But I couldn’t open the box or tear it apart. I tried over and over. I failed at every attempt.

I woke up exhausted and confused about my mental health.

TO
COOK

Berliner

KINO
BIG EYES

HB-YER

4
THIS WOULD SUIT HER

B
YX613H

ANIMALS OF PREY

He drank. She looked away.

The skin was warm against his lips, and its white hairs were glowing in the afternoon sunshine that came through the canopy of the acacia tree. The warm blood seeped slowly from the vein into his mouth. It was thicker than he had expected, and it tasted salty, like rusted iron. His face was smeared with red as he stood up and turned around. He wiped it with the back of his hand, just as he had seen the others do.

It took four minutes to kill the goat. It died quietly in the arms of one of the Maasai. There was no struggle, no kicking, and no shaking. It just went limp. Its eyes fluttered a few times, then remained unshut. Its snout was buried lovingly within the man's broad hands, as if embracing the tenderness with which he took its life.

Where life is at its most essential, death comes with the deepest respect.

We ate the kidneys first, raw and still lukewarm, with a texture like ripe mango. *In the savanna, man is just another predator,* he thought. It eats the organs first. The goat was butchered, grilled, and shared among us all at night. The elders ate the trotters and drank the mead. We all dined, and talked, and sang.

That night, the sprawling savanna shrank. There was the fire and our hosts. The tents shone under the moon. Some young Maasai stood watch.

Because, after all, we are just prey.

KAE 968B

IT WASN'T SHE

However hard he tried, he simply couldn't detect anything desirable about her. He had observed her from behind his book in the coffee corner that she ran at the airport's food court. He still naively believed that books provided excellent cover, but while everyone was scrolling on their smartphones, he was the only one reading—an outcast.

The coffeeshop was designed with the single objective of making money with as little effort as possible. It excelled in the absolute banality that is considered requisite to keep customers the right side of comatose. A single touch of the girl's painted fingernails produced a lifeless coffee from a German stainless steel machine. Customers brought everything else to her, even cups of hot soup, from self-service displays that surrounded the cash register. She spent her day sullenly ringing the till.

Perfect for the job, she excelled at emotionless mediocrity; every gesture was stripped to a minimum, and smiles seemed to require an effort she wasn't paid for.

At two sharp, she stopped working and rested a laminated sheet of paper against the cash register. She locked the glass-door fridges and the soup cauldron, took a sandwich from the counter, and made a cup of tea. She ate her lunch from the windowsill, fully absorbed in her smartphone, while an ongoing stream of customers pulled at the doors of the fridges, fiddled to open the soup warmer, looked helplessly for assistance, and gave up.

She must have worn her high-top black Converse to work for the first time. As soon as she had finished her lunch, she pulled out the laces and put them back, outside in, making sure that they weren't twisted and that both ends were exactly the same length throughout the process. She tied them and tugged the knot underneath the tongue.

As he admired her expertly laced sneakers, he realized that he had never managed to properly tie a one-loop knot and was still doing his laces in two loops—sticking his tongue between his teeth in utmost concentration like any six-year-old—and understood that it wasn't she but her job that lacked seduction.

AS EVEN WORSE

BREAKFAST OF CHAMPIONS

You got it wrong. I'm not referring to Vonnegut's satirical novel. Dwayne Hoover's Exit Eleven Pontiac Village is of little interest to me in this case. Neither is General Mills, for that matter. I haven't eaten Wheaties for breakfast (or for any other meal), and I have no ambition at all to do so in the future.

Let's say my taste in breakfast has evolved and matured. Think almond-paste croissants with a cappuccino served in thick, pre-heated porcelain cups and saucers. Or eggs Benedict on a buttery brioche with a glass of Crémant de Bourgogne. We'll keep the champagne for a little later in the day.

But it's peanut butter tartines with bananas, honey, crispy bacon, and chili when I am in America. I can't help it; it's addictive and incredibly easy to prepare.

Ingredients

Freshly baked sourdough bread, sliced

Peanut butter. I grind my own at Whole Foods and use honey-roasted peanuts.

Uncured, smoked, center-cut bacon. I'm a fan of Nature's Rancher Hickory smoked bacon.

Ripe yellow bananas

Honey. I prefer mountain or heather honey, as it pairs perfectly with the smoky flavor of the bacon.

Chili flakes

Recipe

Preheat the oven to 190°C (375 °F).

Line a sheet pan with foil or baking paper, and lay out the bacon strips evenly.

Bake for 15-20 minutes and flip halfway through.

Drain and cool the bacon on paper towels. They'll become crispier as they cool down.

Lightly toast the sourdough tartines; keep them soft inside.

Slice the banana into rounds.

Spread a thick layer of peanut butter on the tartines and top with honey.

Lay out the banana slices on top of the honey and sprinkle generously with chili flakes.

Add 2-3 bacon strips to the top of each tartine.

If it is too early in the morning for wine, I'd suggest a large glass of whole milk, served ice-cold. Into something stronger already? Try a fortified wine such as a Vintage Port or a Banyuls. A late-vintage (Spätlese) Riesling will surely kickstart your day!

PASTURE
RAISED EGGS

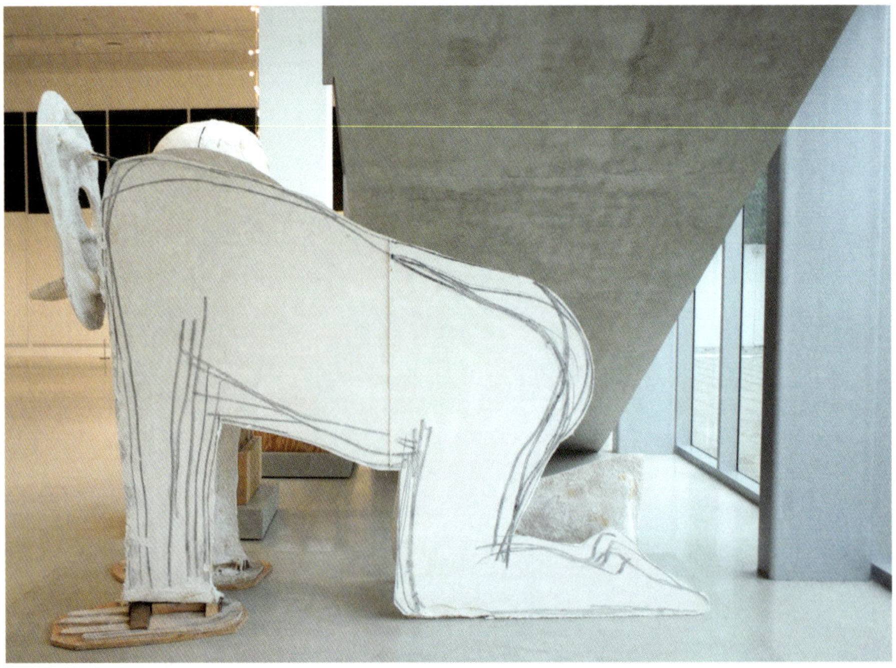

FENDI
FENDI

PROMENADE
des ANGLAIS
CHANTECLER

NO MIDDLE GROUND

I know we drove too fast towards the sea
On muted, twisted roads. You said I should slow down,
Take in the hills, the light, the vineyards (for rosé maybe),
Or drink pastis on squares in godforsaken towns

With older men like me. We saw the rain wash
Salt off buildings, streets, and palms, from clouds in doubt
Of season, shadows long and low. We talked about the blotch
That left you shocked and sore—discarded, as it were, sold out.

Yet little did you know that this was just the start,
That there was pain ahead and death, a sudden surge
Of change that came—in hindsight—hardly in surprise.

"We'll have a kir royal," I told the girl behind the weathered bar
Where artists used to leave their work in lieu. "We want to splurge
And celebrate what comes anew." Because you never compromise.

MERIDIEN

PERFECTLY MISMATCHED

It's in places like these that journeys end before they start anew. Places he never planned to be. On country borders, at highway intersections, or in cities, utterly by accident.

He reached them after dark, driving on quiet country roads or walking on empty sidewalks in towns unexplored. Bars and restaurants were closed for the night. In the early mornings, he always woke up to an out-of-place feeling of ubiquity and total irrelevance.

Fleeting one-off stops that left no traces.

His plane had been canceled, and he found himself in Madrid, tired, cold, hungry, and not wanting to stay in the nondescript hotel the airline had arranged for him. The area was just as plain—a 1970's property developer's paradise. He walked around aimlessly and came across a restaurant that was a few steps down at the back of an eight-story building. With its blue facade, it looked like a dive. He hesitated, but something about it compelled him to enter.

The dining room was loud and full. The floor was painted the same blue as the facade, and the walls were covered with paintings and drawings, hanging crookedly in perfectly mismatched frames. Unquestionably a restaurant where patrons left full, indulged, and completely content.

Tortilla vaga, bocata de papada y trufa, cocochas con almejas en salsa verde, bocaditos de merluza a la romana, ostra escabechada, lentejas.

His dinner was pure taste bud poetry. Memories of Restaurante Sacha had him salivating for over a decade.

They were in town again, some twelve years later. He hadn't told his wife. She found the hotel nondescript. The floor at Sacha was still blue. He'd been drooling over Sacha ever since he knew they'd be back in the city where he once got stranded.

MADRID, 2017

EL GARAJE CIERRA:
DOMINGO A JUEVES 24h.
VIERNES, SÁBADOS VÍSPERAS FESTIVOS 1.30am
ABRIMOS A LAS 6am

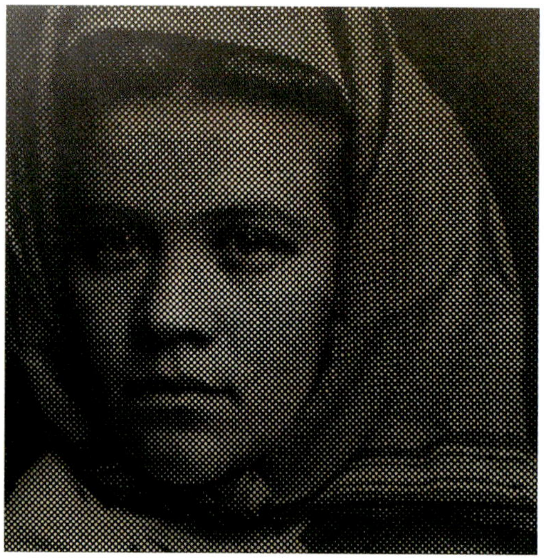

CASTLES MADE OF SAND

I had never expected to meet a man like him in that wind-swept bar with its flimsy plastic tables. He was tall, at least six feet five, and moved slowly. His sand-colored, thin-ribbed corduroy pants hung loosely, and his short-sleeved shirt billowed, overly wide at the arms and neck. It seemed as though there wasn't enough flesh to properly fill the stylish clothes he was wearing.

The place seemed cramped for a man of his stature, with tables packed too closely and stools too low and uncomfortable. A companion gripped his arm tightly as he lowered himself next to me onto the bench in the corner. She struggled to keep him steady.

"I'm so sorry," he said with a hint of self-mockery. "I am a bit unstable lately."

I wasn't sure if he referred to the bar or to his poise.

He spoke articulately. Underneath the gentleness was the natural authority of a man who had been in charge. He wasn't anymore; there were age spots on his hands and deep lines across his ashen face. He must have been in his late seventies, I guessed—a final trip down memory lane for him. For her, it was a pledge.

The variety of chairs wasn't eclectic. The paintings weren't art. The Jimi Café was a nostalgic dive, where blistered azure walls served as the backdrop for DayGlo Hendrix portraits produced by artists whose enthusiasm was far more compelling than their proficiency. It was the watering hole for vacationers on short camel or horseback tours who wore their tie-dyed T-shirts and their hippie crochet tops for the occasion.

He must have been there before.

"It's all an urban legend," he mumbled. She didn't listen. The waiter brought tea.

"*Castles Made of Sand* was written long before Jimmy Hendrix arrived in Essaouira. It was released a year before he got here. Jimmy probably didn't even remember the place; the stuff he smoked was pretty potent."

I pictured him around 1968 as an ebullient young graduate who hit the hippie trail with friends in a second-hand Renault 4 that they drove through France and Spain to board the Tanger ferry in Algeciras; freedom, adventure, and weed.

He was shaking as he held the small glass with both hands to drink his tea. *And so castles made of sand fall into the sea eventually*, Jimmy sang, and his guitar gently wept.

LOSING MY ELEVATION

Every entry into Comedor Mimi is grand. The door opens directly into the dining room, with its four tables and twenty-three chairs. You only get the barest of discretion if River Plate or Boca Juniors score at the precise moment that you walk in. But even then, stepping through the doorway next to the 55-inch television screen with its sound permanently turned off, you'll get a thorough look-over by the 15 or so railroad workers who dine in silence.

At Tolar Grande, sound is a scarce commodity.

Strolling just a few hundred meters had left us feeling cold and disconnected. The altitude had drained not only our bodies but also our perception of embellishments to such a level that we were delighted to be seated at a table with the ugliest of Christmas-themed tablecloths featuring mismatched reindeer, Santa Clauses, and pine cones, even though we were already in March. The tablecloths at the other tables were conspicuously less inspiring. But at thirteen thousand feet above sea level, functionality outranks décor.

We drove for hours, but didn't walk far. The abandoned sites, remnants of an era when there was budget for Argentina's ambitions, emerged amid otherworldly environments that were lifeless and brimming with pink nuances of utter silence—nature at its most imposing and inanimate. Our journey was both the reward of a determination to satisfy an insatiable curiosity and a humbling encounter with vulnerability.

First comes breathlessness; every cell screams for oxygen, and every single movement requires intense effort. It's like you're walking uphill, even though the ground is perfectly flat. Next, it feels like your brain swells against your skull that refuses to expand. Follows dizziness and nausea to end up totally incapacitated, retching on an empty stomach while battling a splitting headache.

That's the price to pay for gazing upon ominous beauty.

HOSTERIA
El Portal
RESTAURANT
Tel.: (0387)155178854

COSTA MEDITERRANEA

SPEEDO SPREZZATURA

The Costa Mediterranea looked as if it was going to swallow the whole town, but curiously, it wasn't out of place in a city built on maritime trade. Trieste is all about the sea. The air is heavy with salt, Drogheria Toso's shop window displays natural sponges, and the Triestini lounge and chat under the pine trees alongside a lengthy stretch of concrete they call the beach.

We hadn't intended to get drunk that day. I know of people who plan getting legless in advance. We don't. A proper bottle of wine over dinner with friends does it for us. We often skip dinner and go out for lunch. With wine, of course. That's about the limit of our recklessness—pretty boring.

What is more epicurean than a delicious lunch at a restaurant chosen with care, savoring the stolen hours when you feel like you should be working but you aren't? A tinge of guilt sharpens the senses. Everything looks fresher in daylight, the food tastes better, and—at the right restaurant—you absorb the animated vibe of business and its expectation of success.

Granzevola a la Triestina needed explanations. The spider crab was cooked to peak flavor at Menarosti, a small trattoria that locals had advised. A bottle of Dut'Un matched it to perfection. But it was the Poli Grappa that graciously appeared at our table after lunch that did it. We had to take a walk along the beach.

There is a perfect term for it in French: *moule-bite*. Thankfully, the French never experienced the Victorian era. Their language remains descriptive and creative, especially when it involves human reproductive anatomy. By comparison, all descriptions of male genitals in tight-fitting swimming trunks seem rude in English, except maybe for budgy smugglers. But who in Italy wants to be compared to a raucous Aussie? Trieste commands elegance. However you want to describe the sunbathers that afternoon, you are hopelessly out of fashion wearing boardshorts on this beach. You are also conspicuous when you are under sixty.

Azure or fiery red, over skinny, hairless legs, or beneath protruding bellies, Triestini wear their minuscule speedos with pride and confidence to get the maximum tan and the most casual upward gazes from behind the designer sunglasses of Italian grandmothers, stretched out—in purported comfort—on thin bathing towels that barely cushion Topolino's concrete flooring.

La bella figura sometimes hurts.

S

MULHOLLAND DRIVE

Mulholland Drive: a serpentine asphalt ribbon that snakes between the spines of the Hollywood Hills. It's a strip of scorched asphalt throughout the day, smoldering under a sun that lacks pity for the wanderers below. By night, it morphs into a dark and intriguing route lit up by the distant light of a city that's been constructed on champagne, despair, celebration, and cheap wine.

Los Angeles unfolds beneath you with a million shimmering lights, each one with its own story, each one belonging to a life of damaged dreams and new beginnings. The view is intoxicating. Not the sort of intoxication you get from overly strong gin-tonics that you prepare in airport lounges; it's the one that reminds you that despite it all, beauty still abounds.

No other city radiates more hope. No other road enables you to revel in such a sense of sublime supremacy.

Mulholland Drive is more than just a road. It's a lifeline and a statement. It's where you escape the everyday harshness for a feeling of affection that you cautiously design and hold onto because it makes you, momentarily, infinite and vulnerable at once.

And so you sit and inhale the city. Each breath fills you with a renewed sense of resolution. As you rise above the chaos to find comfort in the beauty all around you, you feel both mesmerized and numb.

DRY CLEANING
SPECIAL
SUITS $14.99
DRY CLEANERS

SAME DAY
service
CLEANERS

Santa

VENICE
SURF & SUP
LESSONS
RENTALS

NEVER

5

AR WASH
CAR WASH
BIKE LANE

8454

633

LIDO EXCELSIOR

Like scissors cutting through the finest cloth, the Riva ripped through the water of the perfectly flat Laguna, its horizon blurry in the mid-afternoon heat. In front of the varnished mahogany bow, the Lido appeared in all of its shimmering pastels, understated splendor, and expectations of a lazy afternoon on the beach with the kids. She left the selfie-obsessed tourists behind to soak up the afternoon sun and swim in the Adriatic.

The Lido always threw her back into the 1950s with its liberty houses, stately hotels, and immaculate beach tents, regimented along the shore. She admired the staff—men and women—that looked upon their job as a profession, that are proud of their work, every single gesture efficient, precise, and elegant. Nothing seemed to have changed over the last 70 years; its muted beauty was left untouched by progress.

As the sun began to dip lower in the sky, the children gathered for a last game of beach football, imagining Juve, the Azzurri, or the Rossoneri. From the sidelines, mothers observed, ostensibly unmoved but flushed with pride as each side played to win.

At dusk, the greengrocer on the square trimmed her artichokes with a dexterity that left her perplexed and afraid he'd lose his fingers. She knew the children would be famished, so she had planned Carciofi a la Romana for dinner. And she made her Lido afternoon stretch well into the evening.

ANY PART OF THE PIER

GRANDE ALBE

CITY
OF
MARFA

RATTLESNAKES AND SOLITUDE

"I wanted to be more detached," she said.

"The city was slowly choking me. The expanse there was three blocks long and four lanes wide at best. None if you have one of these double-decker tourist buses ahead of you. The open-top ones, you know, the ones where tourists get soaked from taking selfies in the rain. Idiots."

He had often wondered why people dress up as tourists when they are on vacation. Did they studiously shop for hideous t-shirts and capri pants before they left?

"I came here for the endlessness of the horizon, for the pitch-black nights with only stars to guide me, and for the rattlesnakes in the tall grass. Somehow, their closeness soothes me. The proximity of death makes me feel alive again," she added after a brief pause.

Nothing about her was Texan. Except for her tan, maybe. She was slender and well-dressed; half of her white linen shirt was carefully untucked from her designer jeans. Her hair was platinum gray and cut in a bob.

She poured herself another glass from the pitcher of Bud that they shared. The roar of a heavily loaded Freightliner pulling away onto the US-90 numbed what she was saying.

"It will simply be too late," was all he heard. She stood up and hugged the man who just walked in, built like a double-decker bus.

Space is for the farsighted, he realized, and left.

TATE
IES
ANTEED!

Marfa

PRADA
MARFA
PRADA
MARFA
PRADA
PRADA

Carmen's Cafe

Laundry
CLEANING
AND DELIVERY
SERVICE

EL COSMICO

FORECASTING FOOTWEAR

My weather app is on the home screen of my telephone. It's the second icon from the right on the second row from the top. Not that I am particularly interested in the weather. I don't really care, to be honest. Before I take the dog for a walk, I go outside to find out if it's warm or cold and add a layer if necessary. If I think it might rain, I take an umbrella. I don't mind getting wet. I just dislike drying my clothes while I'm wearing them. What I dislike even more is changing my outfit without proper reasons, and rain isn't one of them. In my opinion, the only justifications for changing during the day are job interviews, going to the beach, or attending a funeral. Beaches and funerals have been in short supply lately. And I have yet to be invited for a proper job interview.

But there are times when I check my weather app at least a dozen times a day. I study the ten-day temperature forecast with intent. Not only in Brussels, my home town, but also in Paris, London, Amsterdam, and Frankfurt. I don't have plans to visit any of those cities; they just turn out to be in the cardinal directions. I want to be prepared for what's coming, that's all.

It happens about every six months: I become maniacal about the weather. It's all got to do with socks. If a man's shoes tell you a great many things about him, his socks—or their absence—will reveal the rest.

I have a drawer full of socks. Some have worn thin at the heel or have a toe peeking through. I wear them anyway. If Sophia Loren's toes can poke through a huge hole in her stockings for Annie Leibovitz's photo, why can't mine?

It's crew socks for winter. I have them in navy, red, and tartan. Nothing else. No-show footies are for summer. Only blokes with more testosterone than elegance wear them in the cold of winter. But bicolored silk and lisle knee socks are my favorites. I get my pleated beige shorts and best oxfords out to wear them on exceptionally crisp days in summer and winter. If I were Scottish, I swear I'd wear a kilt.

There's an app that tells you the flight number, departure, and destination of every plane in the sky. There's a Shazam for bird songs, and there's one that suggests the best time to go pee during a movie. But there is no app telling you what socks to wear. If, like me, you dread changing clothes during the day, you'll detest changing socks. A simple mistake cascades into changing your entire outfit.

Socks are crucial. Twice a year, I restock my drawer of footies for crews, or vice versa.

It's that time of the year now; the pavement is covered in leaves, and I have swapped my rosé for a dram of Islay whisky in the evening. I'll need to make a decision soon that will seriously affect the next six months of my life. So I better check my weather app again and get it right.

DANGER
FLUX
CAPACITOR

CHANEL

Budweiser

ON DRIVING

I am a driver.

I ditched my first car when I was 12. It was a 1962 Land Rover, so I managed to get it out of the ditch before anyone noticed. At 14, I backed up trucks at my family's meat plant, bending a few fenders. Or worse.

They noticed.

I got better with age. At least, that's what I believe. Men typically overrate their driving skills. And their genitals. But that's a different subject.

To me, driving is a sensory affair.

The whir of tires on wet asphalt, the rhythmic padam-padam of expansion joints, and the humming of the engine are meditative. They are the result of simple, repetitive actions performed effortlessly. So are roadside landscapes. They appear without a narrative and dissipate just as easily, leaving nothing behind but the purpose of arriving.

It sets my mind on fire.

The constraint of the highway is my boulevard to imaginative freedom. It's like my brain needs the intuitive activity of driving to become truly creative.

By myself, I analyze the world's latest geopolitical crises, brighter than the New York Times. I create the most innovative business plans and negotiate the most complex deals with investors that are dying to get in. I win an argument with my wife. But not a word is spoken. No records are kept.

All trumped by the palpability of the destination.

RICKY
LEE

SAT 8 ORNAMENT &
SPANKSHAFT
OPEN BREAKFAST
LUNCH& DINNER
JOSHUA TREE
SALOON & GRILL
GRASS FED BEEF
14 BEERS ON TAP

JOSHUA TREE
SALOON

650

CUBA LIBRE?

The best thing about traveling to Havana was the connection in Toronto. He wasn't particularly fond of airports. If he could recoup all the time he wasted in airports, he'd be able to take a year off, at least. Like a researcher, he always looked for the terminal's best vantage points to observe human behavior in captivity. They are all just lab rats, he thought, while they believe they just embarked on a journey to freedom.

What he loved the most about Toronto Airport was the world's longest and fastest moving walkway. He walked it twice, just for the fun of it: one minute and three seconds to cover the 290 meters. It was ironic how he, too, was running on the rat wheel of his own free will.

He knew that freedom was merely captivity, brilliantly reframed.

He had needed a break from work. Eight months earlier, he had witnessed another remarkable phenomenon of human behavior: his Japanese business partners lost their temper. The ever-so-composed managers had appeared uncharacteristically nervous at the start of the meeting. They lost it completely half an hour in; a lifetime of cultivated constraint went down the drain in minutes. Things got nasty. He was unsure whether to call his legal counsel or a psychologist.

Nevertheless, the night before his departure for Havana, he had signed off on the final version of the agreement that would set him loose. It took the full eight months and 60 pages of legal articles to describe his future freedom. Everyone regained their composure. The lawyers gained financially.

He found no freedom in Havana, only poverty, dilapidated houses, and merchants selling onions from a worn-out tarp at a makeshift market. Even the pretense of reframing complete control as the ultimate goal of collective Cuban national pride seemed to have gone missing. In the stifling summer heat and the crippling humidity of Havana, he found mostly lethargy beneath the beauty of decay. It felt strangely familiar.

Tokyo was just as hot and humid. He was there to sign the agreement. They negotiated for another three days. He blew his top on the fourth day. The deal was sealed that same afternoon in awkward silence.

He regained his freedom and composure. He had no time to waste on lethargy.

VOGUE
Refined and Polished

GENERAL CONDITIONS OF TRADE

It had taken him the best part of thirty years to understand that he was living with a visual thinker.

Her SMS mentioned a narrow gravel road that ran from the sea, and a red door underneath a large tree blooming with pink flowers. *You can smell the tree from the street. Just ring the bell,* it read.

It was dark when he arrived. The taxi dropped him off in front of a building with a black door. There was no tree in bloom. There was no bell, either. Dogs were barking and yanking their chains. The smell of burning coal hung in the hot evening air. No more airline food tonight, he hoped.

"The *door* was black," was how he greeted her. He knew she'd hate him, but he said it anyway. "The *gate* was red." He emphasized the nouns to add effect. The red wooden gate, attached to elaborately carved but weathered gateposts, was only a few meters from the door where a neighbor had found him at a loss. The man even rang the bell on his behalf. A wiry concierge in a dhoti let him into a courtyard. There was a huge bougainvillea at the back.

"Door, gate, what the hell does it matter? You know what red is, don't you?" She replied, piqued, exactly as he had expected. "How was your flight?"

"Uneventful." He was behaving like an ass again.

The exuberance of novel colors, motifs, and contours was colliding in her head. The semantic conventions of SMSs were inadequate to convey such turmoil in words. But his world was verbal; the words *red door* conjured a detailed image of exactly that. Nothing else. He was disoriented when reality didn't match his mental impressions. Doors don't have gateposts. Gates do. It was that simple.

Someone gave him an ice-cold Kingfisher. Under an overgrown pergola, chefs were preparing two barbecues. One was glowing red with hot coal for the chicken kebabs. They were packing the other one with hay. Freshly caught mackerel were wrapped in banana leaves to be smoked in the burning hay. The beer went down with a hiss.

"I accepted you despite your whining and complaining," she acknowledged his feeble attempt at an apology. She was carefully removing the tiny bones from the mackerel, stacking them artfully on one side of her plate. "But I guess I missed the chance to trade you in."

SHANGHAI UNSHOD

"Everything is fine," he said. "He'll be OK." Victor rang from Henri's phone. He slurred; his voice was thicker than usual, as if he had been drinking—or was it a bad connection? It was two a.m. on a Saturday night in Brussels, where he was ringing from. Drinking was more likely.

Getting a call from our son's phone in the middle of the night was unusual. Hearing Victor on the phone was bizarre.

Our family is not big on calling. Let me rephrase this: We're not big on communicating. We broadly know what's going on in each other's lives, and we let everybody get on with it. Sharing experiences is totally optional and mostly happens over a meal and after at least one decent bottle of wine. There are cheaper ways to stay in touch, but none nicer.

"Nothing seems broken," he said, sounding panicky. "He'll be home in a few days."

Disembarking at Shanghai station, we found ourselves trapped in a stampede of Chinese travelers getting back from their New Year's break. The world stopped. So did we. In China, you go with the flow. We didn't. We obstructed the never-ending stream of people rushing to exit the station, ignoring their disapproving looks.

Earlier that week, the taxi driver was typing WeChat messages while he sped into Hangzhou. As we tried to convince him to slow down or stop texting, he swung around indignantly, crossed the entire width of the motorway, and ended up with two wheels on the grass, still speeding. And just this morning, our taxi to the station burned all traffic lights and brushed a woman on a bike who yelled out furiously. At exactly that time, a taxi in Brussels scooped up Henri on a pedestrian crossing, sent him spinning over the car, and dumped him on the asphalt.

"I'm sorry. He lost one of his shoes. I searched everywhere but couldn't find it." Victor was gasping for air now, terrified.

The world slowly started moving again. Henri was discharged from the hospital, and we returned from Shanghai. We took a taxi home and arrived at the same time. Both rides were uneventful. We never found the missing shoe.

SELLING SOULS AND SHADOWS

The city's pink walls are neither a cruel joke nor a sublime fantasy. They supply the vibrant backdrop of the city's pulse of life, its unrelenting throb, which only slows down during the hottest hours of the day when even the most determined babouche vendors sprawl in the air-conditioned backrooms of their stalls and only foreigners venture out in all their naïveté or foolishness.

He, too, was one of them.

The taxi was a welcoming cocoon that muted the cacophony of hawkers peddling curious wares and hustlers spinning their tales, one more fabricated than the other. Its faux leather backseat clung to his pants as he looked back through the rear window, unsure of what lay ahead for him.

The Mercedes was older than he was and had spent the first 15 years of its life racking up over a million miles as a taxi on Düsseldorf's impeccably maintained streets. Now it avoided potholes on Morocco's gravel routes, its engine still humming delicately as if new. He marveled at how German engineers used to build mechanical masterpieces that could be serviced with the barest of infrastructure in any shed along Africa's dust traps and keep them running forever.

Everyone's selling something here: meat from animals slaughtered that morning at a makeshift abattoir, truckloads full of hay, engine oil, spares for a Singer sewing machine, a piece of their soul, a moment of their time, or just any trinket that may be of use. He loved this mercantilism, stripped of premeditated marketing. It centered solely on immediate, concrete needs—material or fictitious. Each deal was personal, a result of passionate negotiation, a convention rather than a necessity.

At nightfall, darkness paints the sky with streaks of gold and mauve. A transformation takes place, and the allure of shadows starts to play tricks on the walls. This is when the money is counted, commissions are paid out in cash, and tea is poured into opaque glasses to sweeten deals that didn't always go as planned.

Marrakech didn't try to impress him; it just happened naturally. The city is raw and unapologetic; it is messy, chaotic, and sometimes nonsensical. It is poetic in its madness. But only for those who care about the missing verses.

TAXI

مرحبا

31
Râte
Frite
Salade
Thé
Pain
Hassan
31

A PAPER TRAIL OF FELINE EYES

Our house is full of magazines. They are everywhere. Magazines are piled on the floor of our living room. It takes maneuvering to get to the settee; you need the skills of an acrobat to become a couch potato in our home. They are on the stairs. On the landings, they are stacked two rows deep, about 50 high or so. Even our coffee table space is optimized; we put books on top and hoard magazines underneath.

More is more.

Reinhilde knows them inside out. She is equipped with an unmatched visual memory for the relics of an age when information was curated, scarce, valuable, and interesting. She's too committed to throw them out one day. I'm still around as well. Her standards seem to apply to me also, at least as far as her attachment to relics is concerned.

Gisele Bündchen keeps staring at me feverishly from beside the fireplace. Every night, I curtsy to the feline eyes of Adriana Lima on my way to our bedroom. I live with the most beautiful women in the world. They are on the covers of Vogue, Harper's Bazaar, Porter and Vanity Fair. And they are from Brazil.

"It's like taking my own beer to the Munich Beerfest." I joked about our planned family trip to Brazil with my colleagues in the office. Of course, they listened to my fatuousness; I signed their paychecks every month.

My lively imagination has a remarkable capability to transform reality; I'm the eternal optimist. Brazil is ice-cold caipirinhas, superbly elegant women, immaculate beaches, and palm trees. But at 8 a.m. on a Sunday morning, there are blood stains on the sidewalk outside of the Edificio Itália in the center of São Paulo. The last exhausted addicts shuffle towards their hideout for the day. They survived another sleepless night in Crackland.

We have an issue of Quem, a Brazilian tabloid, somewhere in a pile of magazines. We are in it, party-crashers, accidentally photographed during the opening reception of a store at the Iguatemi shopping mall. A huge steel-beamed platform is stuck to the back of the store on the third floor. It's used to land the customized helicopters that they sell inside. They fly low over the favelas, where the waiters who serve the champagne and canapés live.

Screaming headlines surround a stunning woman's portrait on the cover of Quem. I didn't see anyone like her. But I guess reality is for spoilsports.

SHABBY STABLES, HIPPIE CHIC

He was ten or so when he first fell in love. There have been many girls and women since.

He once fell in love quite easily. There were flings that lasted half an hour, sparked by a passing glance, perhaps an attitude, or a fragrance. They burned out just as easily, leaving nothing but a whiff. Some women swept him off his feet for weeks and left him floored. Their images invaded his daily routine, like an incessant pop-up ad that blocked his screen whenever he scrolled. Annoying (yet not unpleasant). Subscribe now! He never did.

But Erika was the first. It always puzzled him why his earliest heartthrob was a buoyant and athletic blonde, when later in life he inevitably swooned over enigmatic and artistic women with dark hair. OK, there was a Debbie Harry lookalike he biked from school with and kissed a few hundred meters from her home (he's never been a great kisser, so she dumped him in no time), but who could blame him for this outlier when Blondie was topping the charts worldwide?

He knew Erika from riding school. Her father was Italian—he owned a furniture factory somewhere in Sicily, he remembered—and her mother was Dutch. She dropped Erika off in a blood-red Alfa Romeo Spider, oozing 1960s hippie chic as she parked in front of the shabby stables and strutted towards the bar, a vibrant splash of color against the carbolineum-infused pinewood stalls.

A baronet owned and ran the manège. He had been discharged from the military (there were rumors of men in leather at stallion nights in the seedy bars of Hanover) and had fallen on hard times. He was tough on us. We'd ride an hour without stirrups, squeezing the saddle with our knees until our adductors were aflame. He'd use the crop on our hands when they weren't together, thumbs up, or make us clean the stables.

Unfortunately, a passion born out of misery doesn't last; Erika just stopped coming. He hadn't even had the guts to talk to her. But he still experienced the anguish of lovesickness. And the solace of filling another wheelbarrow with horse manure.

All that is left of Erika is riding a horse without stirrups. And the red Spider, her mother's Pucci minidress, and the sweet and musky smell of leather soaked in horse sweat.

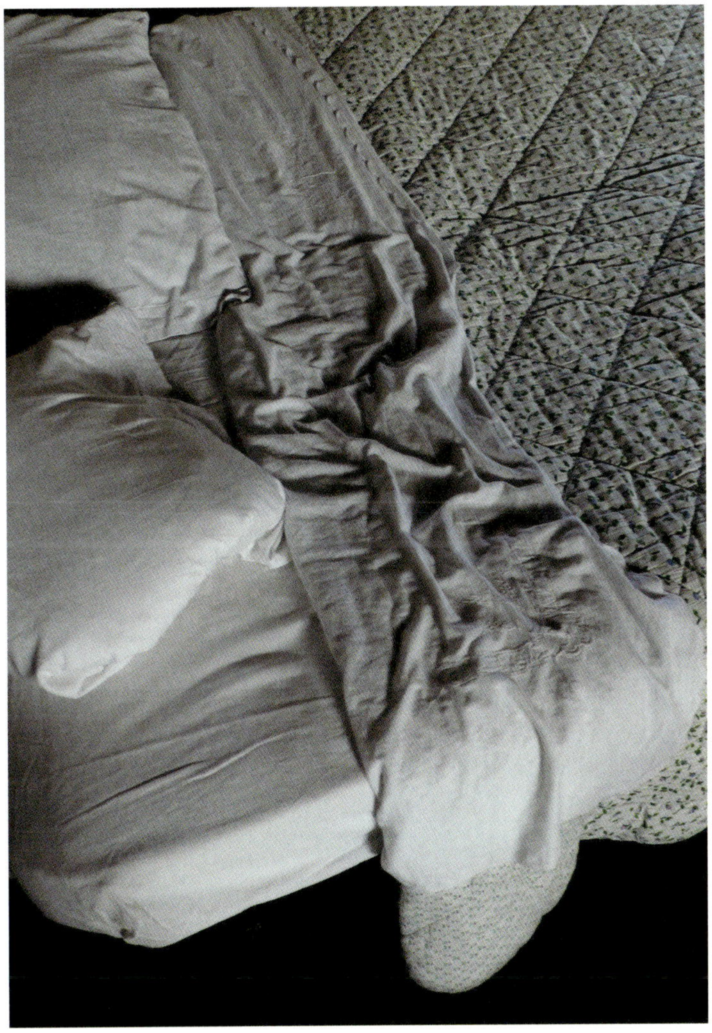

HARDLY A DANCE AT ALL

He marvels at them waltzing life's enchanting tune.
No step misplaced; in perfect youth and sync they swoon
While he dissects the chords. And with a thoughtful face,
His feet respond too late, some kind of clumsy dance in place,

Hardly a dance at all. Life's rhythm nonetheless he learned,
Although its cycles of beginnings often ceased disturbed
With errors, rubble, and concern, he was the artful pioneer
That turned each mistake into a stepping stone for his career.

But as the years advance and doubts cast on previous days,
His past choices come to haunt and today's decisions will enslave—
He knows—what once were fleeting failures, now is the burden that he weighs.

So let the young reclaim, he thinks. They may not crave
For the mistakes he made, nor learn. Yet they will dance and blaze
Through life and fade with little chance to misbehave.

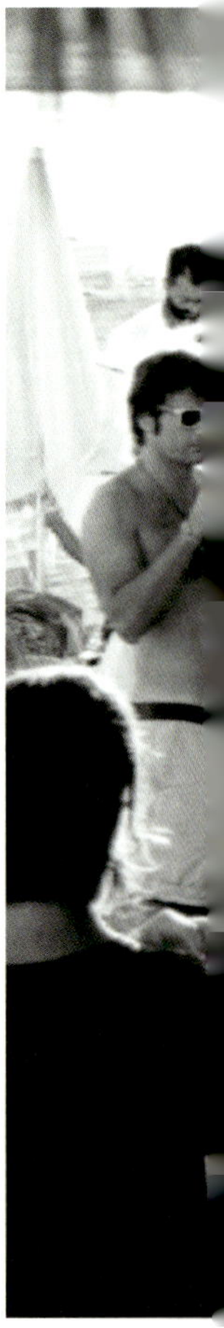

DAYS OF DISCONNECT

They stood out like a sore thumb. Both were dressed identically: Converse sneakers, blue jeans, tailor-made white shirts, and a gray cashmere sweater that hung over their shoulders. And as if that weren't enough, they were the only Europeans on board the 5 p.m. ferry that arrived at the port of Takamatsu. At just 14 and 16, they towered over the other passengers. The breeze tousled their blond hair as they walked off the ramp onto the quay, their faces like thunder.

Earlier in the day, we had left them behind on Naoshima Island. We told them to sort everything out by themselves and catch up with us on Shikoku once they were done. Despite everything, we behaved like respectable parents; we gave them money for lunch and told them where to catch the ferry. They saw things differently. They had parenticide on their minds.

Our Japanese business partners had looked at us in stunned silence when we told them we'd spend a week on Shikoku and the islands in the Seto Inland Sea. Anything beyond the hot springs of Hakone or Fuji-san was considered exotic. We were expected to spend time shopping in Tokyo or discover whatever was left of ancient Japan in Kyoto, then return home. Naoshima simply wasn't on their map, let alone Higashi-iya, deep in the mountains of Shikoku. This was an unconnected Japan, far from any Shinkansen or mobile phone access. A Japan that—embarrassingly—might not meet our expectations. But it did.

Still mesmerized by James Turrell's Backside of the Moon, we noticed that Henri's backpack was missing. We had spent 15 minutes at the Minamidera House to get over the ruthless impact of total darkness and serenely—as if bewitched—witness a single shape emerging out of the black. For a moment, the banality of tangible objects existed in a universe we weren't part of. But missing passports and airline tickets booted us right back into the harshness of reality.

Backpacks don't get lost in Japan. They are found, cared for, and respectfully returned to the owner. Even if they don't speak a word of Japanese, adolescents don't get lost on Naoshima either. That's what we assumed.

As the two of us were visiting Isame Noguchi's atelier in Takamatsu that afternoon, we were no longer so sure if we had acted responsibly. They were not among the first to leave the ship. Neither of us said a word.

The feeling of parenticide subsided. The event became one of the hero stories they share with friends: "When we were kids, our parents abandoned us on an island in Japan." Not the worst opening gambit.

Once their audience is hooked, they explain how they found the backpack, counted their money, and spent it on an extraordinary lunch of somen noodles. While they were sitting cross-legged on a tatami at the best restaurant of the island, we were eating our bento meal in utter silence, freaking out.

NAOSHI A ST DARD²
NAOSHIMA TAND RD²

THE PRICE OF PINK

The beach ball—one of these inflating gadgets with a security valve that put his coordination skills to the test because biting and blowing simultaneously weren't part of his daily routine—was wet with dew. He always felt clumsy and ridiculous when he inflated beach toys like this. Others came prepared with foot pumps. They were halfway through the first chapter in their book, comfortably stretched out on their air mattresses, by the time he stopped blowing and was gasping for breath, his face cherry-red with effort.

He didn't have a clue how the beach ball ended up in his orchard. He was sure that no one of his neighbors played with beach toys in the middle of winter. Nor in summer. The elderly farmers of his village worked on small plots with no more than twenty peach trees. They were painstakingly taken care of, year after year, to produce the most delicate golden peaches, free of spots and bruises, sold as trophies to the wives of corporate salarymen in the food halls of opulent department stores in Kyoto and beyond.

The color had startled him. A vibrant yet delicate pink that was almost translucent. Even if it was made of plastic in a Chinese factory on the outskirts of Shantou, it looked alive—a huge, perfectly curved peach you'd die for. This ordinary globe of air, contained within its thin coating of colored polymer, on the grass near the trunk of a whitewashed peach tree, was the most exquisite image he had ever seen.

He grew the best fruit in the village—peaches that easily fetched eight thousand yen at Isetan and provoked endless oohs and aahs from women in skirt suits and raincoats, their small designer handbags hanging loosely from their left forearm, their mobile phone on their palms in a pose that looked both elegant and uncomfortable.

He daydreamed about them as he meticulously pruned away the upward-growing buds in spring, trimmed off imperfect fruit, and only retained the very best peaches. He'd wrap them in paper to protect them from the sun and carefully harvest them in summer, one by one.

He took the beach ball home. He decided to grow peaches just as pink, sell each one for ten thousand yen, and return to Kyoto solely to bask in oohs and aahs.

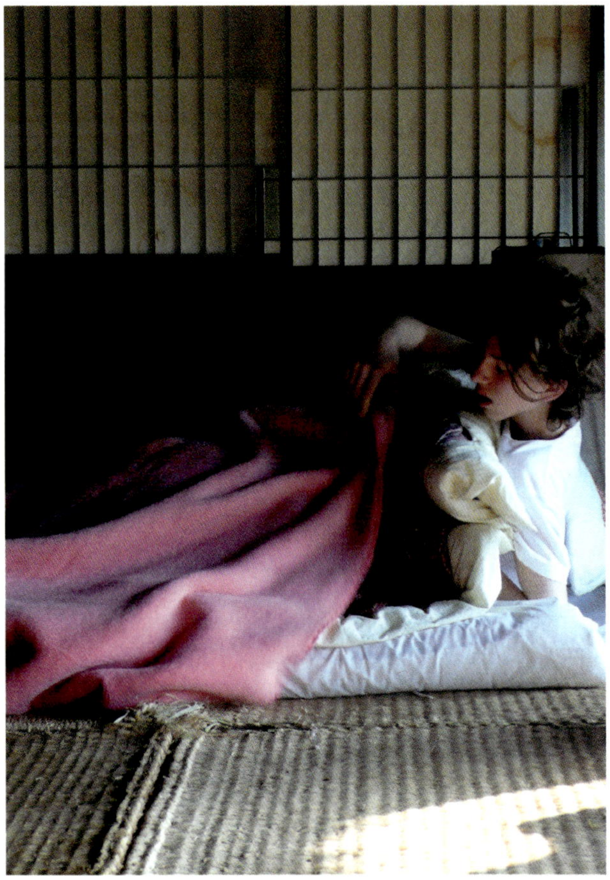

ON LIVING

I have never been a big fan of the masses, but as I walked out of the front door that early summer morning, carrying an overnight bag that contained two pairs of pyjamas bought for the occasion, some books and toiletries, and leaving my watch and wallet behind, my life had been narrowed to one single ambition: to be part of the 97% majority. Because if I had been among the 3 percent minority, it would have marked the last time I walked out of our home.

Statistics are relative. Surgical mortality rates aren't just numbers if it is your body that the surgeons are planning to blunt their scalpels on. They become an obsession—one you can't repress or share with the people around you because you must appear confident.

Well, here I am, part of the 97% survivors, and cheerfully typing this text while an advanced piece of pyrolytic carbon audibly clicks shut about 60 times a minute—a simple activity my own aortic valve ceased to perform some 18 years ago.

I don't use drugs. I don't want to hand over control to synthetic substances manufactured by people whose motivation lies in crafting devastating dependencies. But if the effect of these substances comes even remotely close to the euphoria that I experienced after the operation, I'd be hooked for life.

It was fantastic. I radiated as though I were matching the sun. It shone low into my hospital room on the 8th floor as I woke up early, around five, on top of the world. I was lying in bed with the power of a wet rag, but my brain was working in overdrive while the sterile room radiated with color. I was so alive that I felt immortal.

But it didn't last. Euphoria gave way to irritation. Optimism changed to aggression. Self-awareness turned into selfishness. I was razor-sharp; I became obnoxious. Life with me was ugly for months. Amazingly, my family, friends, and colleagues held on. I owe them big time.

I am no longer self-centered. My obsession with mortality statistics abated. I know that, in the long run, the odds are stacked against me. So what? I live. Every day.

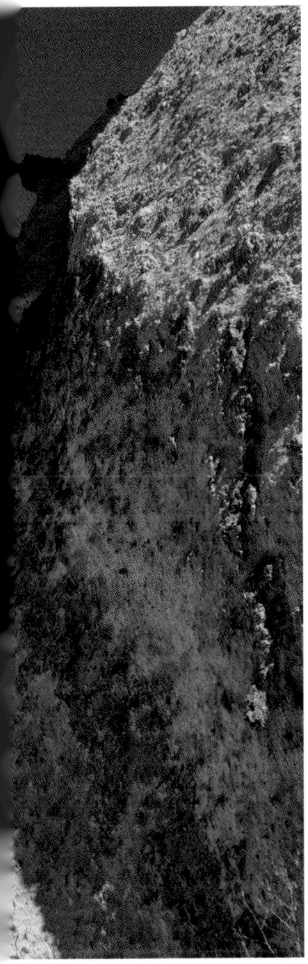

TAMING THE TILL

He enjoyed counting money twice a year. On all other days, it was a senseless, boring activity. But on the 24th and the 31st of December, it gave him a kick. Simply because there was so much of it.

From nine a.m. onwards on these days, customers queued obediently to pick up the orders that they had called in. Most of them were men. Some carried the small printed catalog they were sent earlier that month. The best organized households stood out: they had post-its sticking out of the pages with the ordered deserts in one color, and post-its of another color for the last-minute things that they still needed.

Their wives decided, and they paid. Mostly in cash. Sometimes you could see a glimmer of despair in the men's eyes when they saw the invoice, printed on pristine cotton paper stationery with red painted edges. Their spouses had a taste for luxury. He was eager to oblige.

You can't count money slowly, but you must count it methodically.

He was alone in the office at last. His desk was clean—nothing but his computer and the last of the champagne shared with the team, the neck of the bottle razor-sharp from the sabrage. Tiny bubbles rose to the top of the coupe. He never drank champagne from a flute. And of course, the pile of bills that needed sorting, stacking, and counting. He started to unwind for the first time in months. When the opening and closing balances matched against the day's sales records, he had finished his last job of the year and was ready to go. The bottle was empty, too. He had created order out of chaos, at least temporarily.

The airport was deserted when he got there. The last fifteen flights of the year were about to take off, half-full at best. At midnight, they were flying over Poland. He could see the fireworks below. Lufthansa didn't saber any champagne. He wondered how his family had been celebrating in Cambodia without him.

He arrived at Siem Reap at dusk on New Year's Day. He converted a few hundred dollars into riel at one of the cramped cubicles at the airport. The girl behind the counter smiled shyly when she gave him the thick wad of heavily worn banknotes. They were sorted and neatly stacked face-up.

He didn't count the money. There's always next year.

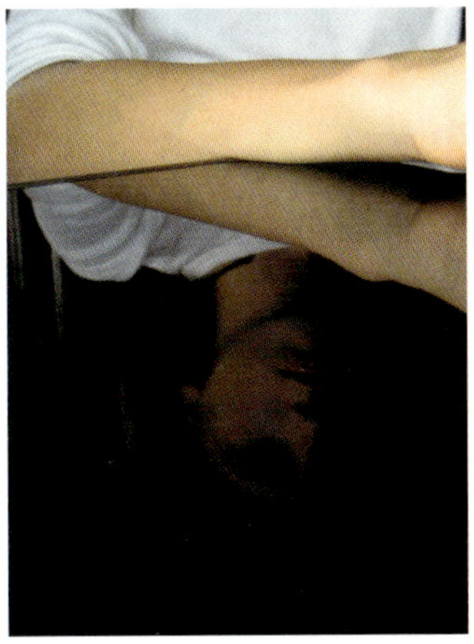

LTD.

HOME | WAULSORT

THREE DOG NIGHTS
(FOR JAMES)

Allow me to introduce myself. I'm James, of noble Jack Russell offspring, chocolate brown spots, 5.5 KG curly tail included and pretty smart. Modest as well, of course. My job is to manage a couple here in Belgium, my home country. They're my masters, at least that's what they think. I walk them regularly because they need some looking after.

Those who had the privilege of meeting me, know that I'm a bit nerdy. Always have been. I love that things are going my way. I chose my family well: they mostly let me get on with my plans for them.

Unfortunately, my master is an early riser. It's still dark when he starts to move around in my bed. Not what I'm prepared for when I'm nicely curled up between his knees, but OK, it's too early for a conflict. Time for a nice stretch—Dachshund style—against his warm body. Bet you, he'll put his phone down and give me a rub. The man has got his priorities right. That's what I like about him.

Ooooh, that's nice. She's awake too. Leaning over for a two-handed rub. Totally oblivious to my master who is in between us. He gets nothing. Better stretch out further. Lift my hind leg a bit. Urgently need a tummy rub. All the way down, please. Life's a bitch…

Get up guys! I want to get out of the bedroom and see what's going on outside. Full speed into the bathroom. I hardly touch the stairs. I'm flying. Hope they won't forget to let me on the balcony for my morning inspection from up above. Like a four-legged, tail-wagging spy drone. Bloody cats still around in the back? Strutting their stuff on the garden walls? Like they friggin' well own the place. I hate them. Open that bathroom door for heaven's sake. Good man. Weather's terrible though. Won't stay out on that balcony for long today. Go back to where it's warm. A couple of woofs and they'll get the picture. They always do. She's letting me in. Hope she has put one of her cashmere sweaters on the chair so that I can curl up a bit more while they are doing their bathroom stuff.

That's it, he tossed his boxer shorts towards me. Smells sooooo good. I know, it's sick, but hey, I'm a dog, right. I live by my nose. Better take it downstairs immediately. Shouldn't trip on his boxers and end up flat on my face at the bottom of the stairs. Don't want the vets to cut me up again like last time. They got my leg working fine, but boy, did it hurt. They shaved all the hair off. My leg looked like one of those hairless cats. With a big zipper on it.

Time to inspect the garden up close. Got my track carved out now. Can't understand why the hell they put a hedge labyrinth in the garden when they could have kept it nice and grassy. But you don't hear me complain. I know the place by heart. Can race it eyes-closed at full speed. Well, slightly slower now; the leg, you know. Must have frightened the cats shitless. None to be seen. Better get in again. It's cold.

Yuck. They're eating pea soup with mint and tofu. For breakfast. Can you possibly imagine? They named me James, so why don't they prepare a proper English breakfast? My masters' sons are called Charles and Henri. It's like the English royal family when we're all together. Only we're in Brussels today, not Buckingham Palace, we don't drive Bentleys and we don't have weird looks and floppy ears. OK, mine are, but I can move them around any way I like. Haven't seen Prince Charles do so recently. Don't think he eats pea soup for breakfast either. Drives a Bentley, though. That would be my car! It would turn the neighbors' two Chihuahuas cactus-green with envy. Boy, do we completely ignore each other!

Morning stroll, morning stroll. Bladder nearly exploding. Need my leash. Where's my leash? It's me who's walking you. Come on, give me my leash and I'll drag you onto the pavement, growling ferociously and commanding respect while I'm parading my turf. I'm the head of the hood. Make way you plump ridgeback from Denmark, piss off, you snub-faced French bull terrier and move over, you stupid golden. This is Jack Russell territory and it's mine! Drups, love you, old girl. And let me have a quick sniff of your backside while I'm at it .

My mistress is in fashion. Holy Pooch in Heaven, that's the best profession any dog can wish for. Real fashion, that is. Like cashmere. Or mohair. Or the most delicate of down lined with silk. Featherweight warmth that she puts on the chair next to hers when she's working. All for me. I've become a textile expert; give me a piece of cashmere and I'll turn it into a perfectly snug nest for my elegant little self within a minute. Dog hair as a bonus.

So what else is there to know about me? Please pick me up when you people are talking so that I can properly take part in your conversation. It's lonely down here and, trust me, your ankles aren't pretty. My leg put an early end to my soccer career. And, oh yes, come and see my rubber fish collection. I play with them every day between 6:30 and 7:30 p.m. sharp. That's right. I told you that I like my days well organized.

Time for bed now. It has been another long day looking after my family. I'm knackered. They'll carry me up the stairs. That's what I trained them for. Wasn't easy: I have been practicing my helpless puppy look for ages. I'll crawl my way in between his knees and lick his thighs. I want to rest my pretty head where it's clean.

Must make a mental note to train him for an evening shower. Gotta start first thing tomorrow when we're off to Waulsort.

James looked after us from Christmas 2009 until February 1st, 2019. Mr. Watson took over on May 22nd, 2019.

AFF
ICH
ES

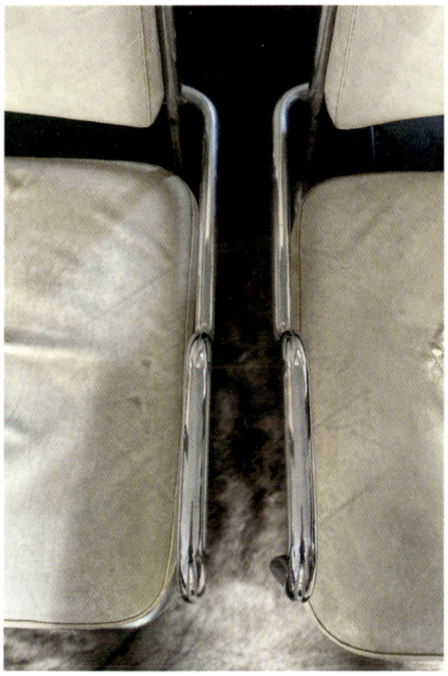

AFF
ICH
ES

AFF
ICH

ABCDEFG
HIJKLM
NOPQRS
TUVWX
YZ

Reinhilde Gielen and Hans Pauwels explore the world in search of fascinating narratives behind concealed beauty. They create true stories about real people, real places, and real companies. Not just stories that stick, but stories that people lose themselves in because they convey timeless values.

As *Aesthetic Nomads*, Reinhilde and Hans work together as a creative duo for content and design. They collaborate closely with companies, organizations, and regions to create dynamic identities through voice, imagery, and storytelling. The brands they value and assist invariably endorse authenticity, tradition, and elegance.

Reinhilde Gielen is an accomplished photographer known for her captivating portrayals of everyday beauty. With a lifelong background as a designer and creative director for luxury fashion, food, beauty, and lifestyle brands, Reinhilde brings a unique perspective to her photographic work. Based in Brussels, she shares her home with her husband, Hans, and their dog, Mr. Watson. Reinhilde's passion for exploration has taken her to unique destinations around the world. As an aesthetic nomad, she spends several months each year immersed in different cultures, capturing intriguing images that exude subdued elegance and sophistication. With her expertly trained eye for distinctive color combinations and light, Reinhilde's photographs are masterful compositions of everyday moments that showcase her artistic vision.

Hans Pauwels stumbled into the enchanting world of books as a kid, and never found his way out again. As the founder and CEO of multiple companies, he mainly wrote business plans, board reports, and sales pitches. And maybe a few wicked but well-versed emails. With newfound freedom, Hans picked up creative writing from where he left off at university. His writing may have matured, but it continues to be pointed, tongue-in-cheek, and tending to the absurd. What else can you expect from a Belgian? Hans lives in Brussels with Reinhilde, his wife, and Mr. Watson, their dog. Together, they travel the world, working as aesthetic nomads.

Follow us at aestheticnomads.com

VIGNETTES

A vast array of people, places, buildings, artworks, artifacts, designs, and images have served as inspiration for Reinhilde's body of work. We have credited all those that we were able to identify in the list provided below. The publisher apologizes for any errors or omissions and welcomes any corrections that should be incorporated in future reprints or editions of this book.

BRUSSELS, Belgium, pp. 14-33: Wirtz International Landscape Architects for all garden architecture; artworks from our private collection by Julia Hetta, Saâdane Afif, Erik Van Lieshout, Carlos Aires, Fred Bervoets, Lars Tunbjörk, Guido Vrolix, Reinhilde Gielen, David Bade, Corinne von Lebusa, Johannes Daniel, Pierre Alechinski, David Bade, Charlotte Schleiffert, Michael Kirkham, Nikita Kadan - CAP FERRET, France, pp. 36-41: Café Ha(a)itza (Pyla sur Mer), Restaurant Chez Hortense, Chez Boulan Restaurant - MÉRIDA, Mexico, pp. 50-71: La Tropical, El Remate Restaurant, Hotel Casa Puuc, Quinta Montes Molina, Montejo 495 Casa Museo, Nuup Colectivo Takto Design, Nectar Restaurant, La Botileria Espirituosa del Sureste - SISAL, Mexico, p. 71: Palmar Club de Playa - COBÁ, Mexico, pp. 72-73: Coqui Coqui Hotel - UXMAL, Mexico, pp. 74-75: Uxmal Archeological site, Mayaland Adventures - TEMOZÓN, Mexico, pp. 76-77: Hacienda Temozón Hotel - ROME, Italy, pp. 78-101: Artworks at the Palazzo dei Conservatori, Musei Capitolini; work of Elisabetta Benassi at the Maxxi; the Galleria Borromini at the Palazzo Spada; works of Alberto Viani, Ettore Spalletti, and Alfonso Balzico at La Galleria Nazionale; details from the Fendi and Valentino stores - ATHENS, Greece, pp. 102-111: Works of Nikolaos Vokos and Theodoros Vryzakis at the National Gallery; work of Vincent Van Gogh at the Basil & Elise Goulandris Foundation; various works at the Acropolis Museum; various works at the National Archeological Museum - NEW YORK, USA, pp. 112-125: Walker Hotel (Manhattan) - BEACON, USA, pp. 119-125: Various works of Dan Flavin, Richard Serra, John Chamberlain, and Dorothea Rockburne at the Dia:Beacon - BUENOS AIRES, Argentina, pp. 126-139: Pizzeria Picsa, Restaurante Casa Cavia, Restaurante Farinelli; artworks of Mathias Goeritz and Yente-Del Prete at the Malba - LA PLATA, Argentina, pp. 140-141: Casa Curutchet, Le Corbusier - MILAN, Italy, pp. 142-151: Fondazione Prada, bar Camparino in Galleria, Pasticceria Marchesi 1824, Ristorante A Santa Lucia, Pirelli HangarBicocca - JOSÉ IGNACIO, Uruguay, pp. 152-173: Parador La Huella, Restaurante La Susana, James Turrell, Ta Khut Skyspace at Posada Ayana, La Linda Panadería & Café - CAPE COD, USA, pp. 174-189: Hatch House, Cape Cod Modern House Trust - VENICE, Italy, pp. 190-199: La Biennale di Venezia with works of Anish Kapoor at the Gallerie dell'Accademia & Palazzo Manfrin, Chun Kwang Young at Palazzo Contarini Polignac, Rony Plesl at Chiesa di Santa Maria della Visitazione, Ai Weiwei at Abbazia di San Giorgio Maggiore, Bruce Nauman at Punta della Dogana, Simone Leigh at Giardini, Anselm Kiefer at Palazzo Ducale, and Julian Schnabel at Museo Correr - SHELTER ISLAND, USA, pp. 200-205: Sunset Beach restaurant - THE HAMPTONS, USA, pp. 206-209: Carissa's The Bakery, The Surf Lodge

NOTES

Please note that this photo book is a collection of images capturing moments in both public and private settings. While we have made every effort to respect privacy and individual rights, some individuals depicted in the photos may be recognizable. The inclusion of individuals in the images is purely accidental and not intended for commercial use or to infringe upon their privacy rights.

PARIS, France, pp. 210-223: Anri Sala at Bourse de Commerce; artworks of Dana Lok and Kudzanai-Violet Hwami at PARIS+ par Art Basel; French Communist Party Headquarters designed by Oscar Niemeyer at Espace Niemeyer; Restaurant Mimosa; details of retail stores of Chanel, Dior, Louis Vuitton, Dolce & Gabbana, Dries Van Noten, and Miu Miu; Hotel SO/Paris; artwork of Alicja Kwade at Place Vendôme - BUTE, Scotland, pp. 230-237: Mount Stuart - ARGYLL, Scotland, pp. 238-239: Ardanaiseig - FORTE DEI MARMI, Italy, pp. 250-251: Ristorante Maitó - PALM SPRINGS, USA, p. 265: "There is nothing in the desert, and no man needs nothing", quote from the film Lawrence of Arabia - BERLIN, Germany, pp. 282-291: Alexandra Kehayoglou at Kaufhaus Herzog, Roman Signer at Kindl Center for Contemporary Art, Rudolf Stingl at Neue Nationalgalerie, Johannes Daniel - COTSWOLDS, England, pp. 300-307: Blenheim Palace - MIAMI, USA, pp. 308-315: Artworks of Lynette Yiadom-Boakye and Peter Liversidge at Art Basel Miami, Frank Stella at NSU Art Museum, Pérez Art Museum, Thomas Houscago at De La Cruz Foundation, Ernesto Neto at Margulies Collection, Fendi Store Miami - MADRID, Spain, pp. 322-329: Artworks of Andrea Procaccini and Francisco de Zurbarán at the Prado Museum, artworks of Eduardo Berliner, José Maria Sicilia, Zhang Dali, and Muntean/Rosenblum at ARCO Madrid, artworks of William Trost Richards and Lucian Freud at Museo Nacional Thyssen-Bornemisza, Restaurante Sacha LOS ANGELES, USA, pp. 356-365: Fitzpatrick-Leland House by MAK Center for Art and Architecture, Cindy Sherman at The Broad - LIDO DI VENEZIA, Italy, pp. 366-373: Hotel Excelsior Venice Lido - MARFA, USA, pp. 374-381: Elmgreen & Dragset on U.S. Route 90, Judd Foundation, El Cosmico - LONDON, England, pp. 382-389: Sir Lawrence Alma-Tadema at the Leighton House Museum - JOSHUA TREE, USA, pp. 390-401: Joshua Tree Saloon & Grill, Acido Dorado by Robert Stone Design, ACE Hotel, Parker Hotel -, CHANDIGARH, India, pp. 412-418: Chandigarh Capitol Complex by Le Corbusier - SHILIMB, India, pp. 420-421: Hilton Shillim Estate (under construction) - RIO DE JANEIRO, Brazil, pp. 444-449: Niterói Arts Centre by Oscar Niemeyer, garden of Instituto Moreira Salles by Roberto Burle Marx - PARATY, Brazil, pp. 452-453: Pousada Picinguaba - SÃO PAULO, Brazil, pp. 454-455: Instituto Bardi by Lina Bo Bardi - NAOSHIMA, Japan, pp. 470-473: Chichu Art Museum - TAKAMATSU, Japan, pp. 474-475: The Isamu Noguchi Garden Museum Japan - HIGASHI-IYA, Japan, pp. 482-483: Chiiori - WAULSORT, Belgium, pp. 502-503: Artworks from our private collection, including works by Johan Parmentier and Daniel Kruger

ACKNOWLEDGMENTS

Thank you to Charles-Axel and Henri-Jérôme for your enthusiasm and for encouraging your parents to continue doing crazy things like making this book.

Thank you Renaat De Vos for making sure that our bizarre design ideas also make sense in a digital format.

Thank you Jonathan North for editing the texts and turning my innate absurdism into palatable content for a larger audience.

Thank you Wim Pauwels and the team at Beta-Plus for believing in us and for making Aesthetic Nomads possible.

150 copies of this publication have been signed by the authors and numbered 1/150 to 150/150 in a limited edition with color edging.

Aesthetic Nomads

www.aestheticnomads.com

First edition published by Beta-Plus

Design: Reinhilde Gielen

Beta-Plus Publishing

Avenue Louise 367

B-1050 Brussels

Belgium

www.betaplus.com

ISBN 978-2-87550-135-6

Printed in Belgium

FSC www.fsc.org
MIX
Paper | Supporting responsible forestry
FSC® C014767